Short Textbooks in Logic

Series editors

Fenrong Liu, Department of Philosophy, Tsinghua University, Beijing, China
Hiroakira Ono, School of Information Science, Japan Advanced Inst Sci & Tech, Nomi City, Ishikawa, Japan
Jeremy Seligman, The University of Auckland, Auckland, New Zealand

This book series contains textbooks on various topics in logic. Though each book can be read independently, the series as a whole gives readers a comprehensive view of logic at the present time. Each book in the series is written clearly and concisely, providing plenty of well-planned examples and exercises, and supplying explanations of the scope and motivation of the topics. The topics discussed in the series range from mathematical and philosophical logic to logical methods applied to computer science, artificial intelligence and cognitive science. The series will contain both introductory and advanced texts. The intended readership includes advanced undergraduate and graduate students in philosophy, mathematics, computer science and related fields. The series is also suitable for self-taught learning.

More information about this series at http://www.springer.com/series/15706

Eric Pacuit

Neighborhood Semantics for Modal Logic

 Springer

Eric Pacuit
Department of Philosophy
University of Maryland
College Park, MD
USA

ISSN 2522-5480 ISSN 2522-5499 (electronic)
Short Textbooks in Logic
ISBN 978-3-319-67148-2 ISBN 978-3-319-67149-9 (eBook)
https://doi.org/10.1007/978-3-319-67149-9

Library of Congress Control Number: 2017951183

© Springer International Publishing AG 2017
This work is subject to copyright. All rights are reserved by the Publisher, whether the whole or part of the material is concerned, specifically the rights of translation, reprinting, reuse of illustrations, recitation, broadcasting, reproduction on microfilms or in any other physical way, and transmission or information storage and retrieval, electronic adaptation, computer software, or by similar or dissimilar methodology now known or hereafter developed.
The use of general descriptive names, registered names, trademarks, service marks, etc. in this publication does not imply, even in the absence of a specific statement, that such names are exempt from the relevant protective laws and regulations and therefore free for general use.
The publisher, the authors and the editors are safe to assume that the advice and information in this book are believed to be true and accurate at the date of publication. Neither the publisher nor the authors or the editors give a warranty, express or implied, with respect to the material contained herein or for any errors or omissions that may have been made. The publisher remains neutral with regard to jurisdictional claims in published maps and institutional affiliations.

Printed on acid-free paper

This Springer imprint is published by Springer Nature
The registered company is Springer International Publishing AG
The registered company address is: Gewerbestrasse 11, 6330 Cham, Switzerland

To Lauren and Lily

Preface

Neighborhood models generalize the well-known relational models, or Kripke models, for modal logic. Although the idea underlying neighborhood models is implicit in the seminal work of McKinsey and Tarksi (1944), they were first formally defined by Dana Scott (1970) and Richard Montague (1970). The original motivation for generalizing relational models was to provide semantics for the so-called *classical* systems of modal logic:

> The qualification "classical" has not yet been given an established meaning in connection with modal logic....Clearly one would like to reserve the label "classical" for a category of modal logics which—if possible—is large enough to contain all or most of the systems which for historical or theoretical reasons have come to be regarded as important, and which also posses a high degree of naturalness and homogeneity.
>
> (Segerberg 1971, pg. 1)

This quote is from Krister Segerberg's dissertation, *An Essay in Classical Modal Logic* (1971), that included early results about neighborhood models and the classical systems of modal logic, also called non-normal modal logics, that correspond to them. A few years later, Brian Chellas incorporated these and other salient results in part III of his textbook *Modal Logic: An Introduction* (1980). Nevertheless, in the apparent absence of applications, neighborhood models and non-normal modal logics were studied mainly in view of their intrinsic mathematical interest. A notable exception is David Lewis's seminal book on conditional logic (Lewis 1969). The so-called *Lewis sphere models* are one of the earliest and most interesting applications of neighborhood models.

Interest in neighborhood models increased steadily over the past 30 years. Neighborhood models form an interesting and rich class of mathematical structures that can be fruitfully studied using modal logic. This is most evident in the extensive literature on the topological interpretation of modal logic. In this book, I highlight additional motivations for studying neighborhood models:

- Neighborhood models naturally show up when studying game theory.
- Neighborhood models can be used to represent the evidence and beliefs of a rational (and not so rational) agent.
- Neighborhood models offer an interesting new perspective on the Barcan and Converse Barcan formulas in the first-order modal logic.

Finally, one can learn a great deal about normal systems of modal logic by looking at how these systems behave in a more general semantics.

This book will quickly familiarize the reader with the general theory of neighborhood semantics for modal logic. I explain how neighborhood models fit within the large family of semantic frameworks for modal logic. In addition, I explain both the pitfalls and potential uses of neighborhood models. This book is designed to be a supplemental text for a course on modal logic, logic in AI, or philosophical logic (either at the advanced undergraduate or graduate level), or as a source for researchers interested in using neighborhood structures in their work. One of the constraints on writing a book for this series is that the length must be kept short. This means that I had to make some hard choices about which topics to leave out. Some topics are only briefly discussed, such as proof theory for non-normal modal logics and the topological interpretation of modal logic. Other topics, such as coalgebraic models for modal logic, are not discussed in any detail.

The book is divided into three chapters. Chapter 1 motivates our study by discussing applications of non-normal modal logics and different interpretations of neighborhood models. Chapter 2 is focused on the core logical theory. This includes questions about axiomatizations, definability, decidability, and relationships with other logical systems. Chapter 3 surveys different extensions to the basic logical framework, including the first-order quantifiers, dynamic modalities, and multi-agent modalities. Finally, there is a short Appendix providing some background on relational semantics for modal logic. Exercises are included throughout the text. There is a website for this book:

$$\text{http://pacuit.org/modal/neighborhoods}.$$

The website includes an extended appendix, solutions to some of the exercises, lecture slides and videos on topics discussed in this book, and links to relevant readings and tutorials on modal logic.

This book grew out of notes on neighborhood semantics that I prepared for two week-long ESSLLI (European Summer School for Logic, Language and Information) courses on neighborhood semantics for modal logic: ESSLLI 2007 in Dublin, Ireland, and ESSLLI 2014 in Tübingen, Germany. I thank the students who attended these courses. I am also grateful for the students that attended a short course at the Institute for Logic, Language and Computation at the University of Amsterdam in June 2016. I thank the following people whose comments and discussion over the years have influenced various aspects of this book: Albert Anglberger, Alexandru Baltag, Johan van Benthem, Nick Bezhanishvili, David Fernandez-Duque, Huimin Dong, Mel Fitting, Nic Fillion, Valentin Goranko, Davide Grossi, Helle Hvid Hansen, Wes Holliday, Hanna Karpenko, Marcel Kiel,

Dominik Klein, Clemens Kupke, Fenrong Liu, Minghui Ma, Larry Moss, Rohit Parikh, Olivier Roy, Katsuhiko Sano, Shawn Standefer, Sonja Smets, and Dag Westerståhl. I am very grateful to an anonymous reviewer and to Johan van Benthem who both provided very extensive and insightful comments on an early draft of this book. Both sets of comments led to many significant improvements to this text. Sadly, one person I would like to thank, Horacio Arló-Costa, is no longer with us. Horacio initiated my interest in neighborhood models for modal logic. I miss our discussions about philosophical logic, formal epistemology, and decision theory over bad coffee at the Graduate Center. Finally, I thank my family, especially Lauren and Lily, for their love and support.

College Park, MD, USA
January 2017

Eric Pacuit

Contents

1 **Introduction and Motivation** 1
   1.1 Subset Spaces................................... 1
   1.2 Language and Semantics 4
       1.2.1 Neighborhood Frames and Models 6
       1.2.2 Additional Modal Operators 9
       1.2.3 Reasoning About Subset Spaces 10
   1.3 Why Non-normal Modal Logic?......................... 12
   1.4 Why Neighborhood Structures? 19
       1.4.1 Topological Models 20
       1.4.2 Hypergraphs 24
       1.4.3 Conditional Logic 26
       1.4.4 A Logic of Evidence and Belief 32
       1.4.5 Coalitional Logic............................ 34

2 **Core Theory** 39
   2.1 Expressive Power and Invariance 40
   2.2 Alternative Semantics for Non-normal Modal Logics........... 46
       2.2.1 Relational Models............................ 47
       2.2.2 Generalized Relational Models 49
       2.2.3 Multi-relational Models......................... 51
       2.2.4 Impossible Worlds 51
   2.3 The Landscape of Non-normal Modal Logics................ 52
       2.3.1 A Non-normal Extension of **K** 58
       2.3.2 Completeness 60
       2.3.3 Incompleteness and General Frames 66
   2.4 Computational Issues 69
       2.4.1 Filtrations 69
       2.4.2 Complexity 75
       2.4.3 Proof Theory for Non-normal Modal Logics............ 76
   2.5 Frame Correspondence 80

	2.6	Translations	82
		2.6.1 From Neighborhoods to Orders	83
		2.6.2 The Normal Translation	89
		2.6.3 The Standard Translation	93
3	**Richer Languages**		97
	3.1	Universal Modality and Nominals	97
		3.1.1 Non-normal Modal Logic with the Universal Modality	98
		3.1.2 Characterizing Augmented Frames	102
	3.2	First-Order Neighborhood Structures	104
		3.2.1 Syntax and Semantics	104
		3.2.2 The Barcan and Converse Barcan Schema	106
		3.2.3 Completeness	109
	3.3	Common Belief on Neighborhood Structures	116
	3.4	Dynamics with Neighborhoods: Game Logic	124
	3.5	Dynamics on Neighborhood Structures	130
		3.5.1 Public Announcements	130
		3.5.2 Evidence Dynamics	135

Appendix A: Relational Semantics for Modal Logic 139

References 147

Chapter 1
Introduction and Motivation

Generally speaking, there are two different ways to motivate the study of a logical framework. The first is to identify an interesting class of mathematical structures and to argue that some particular logical system is the best one with which to reason about these structures (or to *describe* interesting aspects of these structures). The second is to motivate interest in certain "patterns of reasoning" and to argue that some logical system naturally represents these patterns. This chapter introduces neighborhood semantics and weak systems of modal logic with both motivations in mind.

Before turning to questions of motivation, I introduce the basic concepts and definitions used throughout this book. Section 1.1 introduces subset spaces. The core focus of this book is on the basic modal language (Sect. 1.2), interpreted on neighborhood models (Sect. 1.2.1). Section 1.2.3 digresses briefly to introduce a bi-modal logic for reasoning about subset spaces.[1] The final two sections are devoted to motivating our study. I start in Sect. 1.3 by briefly discussing a number of weak systems of modal logic. Each of these logical systems is an example of a so-called *non-normal modal logic* (see Sect. 2.3 for a definition of "non-normal modal logic"). However, I want to stress that the main focus of this book is not non-normal modal logics per se, but, rather, neighborhood semantics for modal logic. While the neighborhood models defined in Sect. 1.2.1 do provide a semantics for the modal languages discussed in Sect. 1.3, they are *not* the best choice of semantics for many interpretations of the modal operator. Nonetheless, I argue in Sect. 1.4 that neighborhood models are an interesting class of mathematical structures that can be fruitfully studied using modal logic.[2]

1.1 Subset Spaces

Sets paired with a distinguished collection of subsets are ubiquitous in many areas of mathematics. They show up as *topologies*, *ultrafilters*, or *hypergraphs* (also called *simple games*), to name three of the most usual suspects. For any non-empty set

---

[1] This section can be skipped on a first reading.
[2] Although this text is self-contained, readers that have not studied relational semantics for modal logic should consult Appendix A for a brief introduction.

W, let $\wp(W)$ be the power set of W—i.e., the collection of all subsets of W. A tuple $\langle W, \mathcal{U} \rangle$ where $W \neq \emptyset$ and $\mathcal{U} \subseteq \wp(W)$ is called a **subset space**. Typically, the collection $\mathcal{U} \subseteq \wp(W)$ satisfies certain algebraic properties. I list some salient properties below:

1. $\mathcal{U}$ is **closed under intersections** provided that for any collection of sets $\{X_i\}_{i \in I}$ such that for each $i \in I$, $X_i \in \mathcal{U}$, we have $\bigcap_{i \in I} X_i \in \mathcal{U}$. If $|I| = 2$, then $\mathcal{U}$ is said to be closed under **binary intersections**. If I is finite, then $\mathcal{U}$ is said to be closed under **finite intersections**. More generally, for any cardinal κ, $\mathcal{U}$ is said to be **closed under κ-intersections** (**closed under less than or equal to κ intersections**) provided that for each collection of sets $\{X_i\}_{i \in I}$ from $\mathcal{U}$ with $|I| = \kappa$ ($|I| \leq \kappa$), we have $\bigcap_{i \in I} X_i \in \mathcal{U}$.
2. $\mathcal{U}$ is **closed under unions** provided that for any collection of sets $\{X_i\}_{i \in I}$ such that for each $i \in I$, $X_i \in \mathcal{U}$, we have that $\bigcup_{i \in I} X_i \in \mathcal{U}$. The same comments as above about binary and finite collections apply here, as well: In particular, for any cardinal κ, $\mathcal{U}$ is said to be **closed under less than or equal to κ-unions** provided that for each collection of sets $\{X_i\}_{i \in I}$ from $\mathcal{U}$ with $|I| \leq \kappa$, $\bigcup_{i \in I} X_i \in \mathcal{U}$.
3. $\mathcal{U}$ is **closed under complements** provided that for each $X \subseteq W$, if $X \in \mathcal{U}$, then $X^C \in \mathcal{U}$, where $X^C = \{w \mid w \in W, w \notin X\}$ is the complement of X.
4. $\mathcal{U}$ is **closed under supersets** provided that for each $X \subseteq W$, if $X \in \mathcal{U}$ and $X \subseteq Y \subseteq W$, then $Y \in \mathcal{U}$. In this case, $\mathcal{U}$ is also said to be **monotonic** or **supplemented**.
5. $\mathcal{U}$ is a **clutter** if $\emptyset \notin \mathcal{U}$ and there are no $X, Y \in \mathcal{U}$ such that $X \subset Y$ (I write $X \subset Y$ when X is a strict subset of Y).
6. $\mathcal{U}$ **contains the unit** if $W \in \mathcal{U}$; and $\mathcal{U}$ **contains the empty set** if $\emptyset \in \mathcal{U}$.
7. The set $\bigcap \mathcal{U}$ is called the **core of** $\mathcal{U}$; when $\bigcap \mathcal{U} \in \mathcal{U}$, $\mathcal{U}$ is said to **contain its core**.
8. $\mathcal{U}$ is **proper** if $X \in \mathcal{U}$ implies $X^C \notin \mathcal{U}$.
9. $\mathcal{U}$ is **consistent** if $\emptyset \notin \mathcal{U}$; and $\mathcal{U}$ is **non-trivial** if $\mathcal{U} \neq \emptyset$.

The remainder of this section contains a number of simple observations about subset spaces that will be used throughout the text. I start with a discussion of monotonic subset spaces since they play an important role in this book.

Lemma 1.1 *$\mathcal{U}$ is closed under supersets iff $X \cap Y \in \mathcal{U}$ implies that $X \in \mathcal{U}$ and $Y \in \mathcal{U}$.*

Proof The left to right direction is trivial, as $X \cap Y \subseteq X$ and $X \cap Y \subseteq Y$. For the right to left direction, suppose that for any $X, Y \subseteq W$, if $X \cap Y \in \mathcal{U}$, then $X \in \mathcal{U}$ and $Y \in \mathcal{U}$. Let $Z \subseteq Z' \subseteq W$ and $Z \in \mathcal{U}$. We must show $Z' \in \mathcal{U}$. Since $Z \subseteq Z'$, $Z \cap Z' = Z$, and so we have $Z \cap Z' = Z \in \mathcal{U}$. Hence, $Z \in \mathcal{U}$ and $Z' \in \mathcal{U}$, as desired. $\square$

Definition 1.1 (*Non-monotonic Core*) Suppose that $\mathcal{U}$ is a monotonic collection of subsets of W. The **non-monotonic core**, denoted $\mathcal{U}^{nc}$, is a subset of $\mathcal{U}$ defined as follows:

1.1 Subset Spaces

$$\mathcal{U}^{nc} = \{X \mid X \in \mathcal{U} \text{ and for all } X' \subseteq W, \text{ if } X' \subset X, \text{ then } X' \notin \mathcal{U}\}.$$

The non-monotonic core of $\mathcal{U}$ is the set of minimal elements of $\mathcal{U}$ under the subset relation. It is not hard to see that if $\mathcal{U}$ is a monotonic collection of finite sets, then $\mathcal{U}^{nc} \neq \emptyset$. However, in general, it is not true that every monotonic collection of sets has a non-monotonic core, as the following example illustrates.

Example 1.2 Suppose that $W = (-1, 1) = \{x \mid x \in \mathbb{R} \text{ and } -1 \leq x \leq 1\}$ and $\mathcal{U} = \{X \subseteq (-1, 1) \mid (-\frac{1}{n}, \frac{1}{n}) \subseteq X \text{ for some natural number } n \geq 1\}$. It is not hard to show that $\mathcal{U}$ is monotonic (if $X \in \mathcal{U}$, then there is a natural number $n \geq 1$ such that $(-\frac{1}{n}, \frac{1}{n}) \subseteq X$; hence, if $X \subseteq X'$, then $(-\frac{1}{n}, \frac{1}{n}) \subseteq X'$, and so $X' \in \mathcal{U}$). However, the non-monotonic core of $\mathcal{U}$ is empty—i.e., $\mathcal{U}^{nc} = \emptyset$: Suppose that $X \in \mathcal{U}$. I will show that there must be a set $X' \subseteq X$ such that $X' \in \mathcal{U}$. Since $X \in \mathcal{U}$, there is a natural number n such that $(-\frac{1}{n}, \frac{1}{n}) \subseteq X$. Let $X' = (-\frac{1}{n+1}, \frac{1}{n+1})$. Then, $X' \subseteq (-\frac{1}{n}, \frac{1}{n}) \subseteq X$ and $X' \in \mathcal{U}$. Thus, $X \notin \mathcal{U}^{nc}$. Since X is an arbitrary element of $\mathcal{U}$, it must be the case that $\mathcal{U}^{nc} = \emptyset$.

Suppose that $\mathcal{U}$ is a monotonic collection of finite sets. Then, $\mathcal{U}^{nc} \neq \emptyset$; and, in fact, $\mathcal{U}^{nc}$ completely determines the elements of $\mathcal{U}$.

Definition 1.3 (*Core Complete*) A monotonic collection of sets $\mathcal{U}$ is **core-complete** provided that for all $X \in \mathcal{U}$, there exists a $Y \in \mathcal{U}^{nc}$ such that $Y \subseteq X$.

If $\mathcal{U}$ is core-complete, then every element of $\mathcal{U}$ contains some element of the non-monotonic core (so, in particular, if $\mathcal{U} \neq \emptyset$, then $\mathcal{U}^{nc} \neq \emptyset$). Thus, $\mathcal{U}^{nc}$ represents $\mathcal{U}$ without any redundancies. Furthermore, note that if $\mathcal{U}$ is monotonic and only contains finitely many sets, then it is core-complete.

Returning to the more general setting (in which the collections of sets need not necessarily be monotonic), the following definition lists some well-known subset spaces that will be discussed in this book.

Definition 1.4 Let W be a non-empty set and $\mathcal{U} \subseteq \wp(W)$. Then:

1. $\mathcal{U}$ is a **filter** if $\mathcal{U}$ contains the unit, and is closed under binary intersections and closed under supersets. $\mathcal{U}$ is a proper filter if, in addition, $\mathcal{U}$ does not contain the emptyset.
2. $\mathcal{U}$ is an **ultrafilter** if $\mathcal{U}$ is a proper filter and for each $X \subseteq W$, either $X \in \mathcal{U}$ or $X^C \in \mathcal{U}$.
3. $\mathcal{U}$ is a **topology** if $\mathcal{U}$ contains the unit, the emptyset, and is closed under finite intersections and arbitrary unions.
4. $\mathcal{U}$ is **augmented** if $\mathcal{U}$ contains its core and is closed under supersets.

Now, I consider augmented collections in a bit more detail.

Exercise 1.1 Prove the following:

If $\mathcal{U}$ is augmented, then $\mathcal{U}$ is closed under arbitrary intersections. In fact, if $\mathcal{U}$ is augmented, then $\mathcal{U}$ is a filter.

Of course, the converse of the last statement in the above exercise is false. But that is not very interesting since it is easy to construct collections of sets closed under intersections but not closed under supersets. A much more interesting fact is that there are consistent filters that are not augmented.

Fact 1.5 *There are consistent filters that are not augmented.*

Proof The collection of sets from Example 1.2 is an example of a consistent filter that is not augmented: Recall that $W = (-1, 1) = \{x \mid x \in \mathbb{R} \text{ and } -1 \leq x \leq 1\}$, and $\mathcal{U} = \{X \subseteq (-1, 1) \mid (-\frac{1}{n}, \frac{1}{n}) \subseteq X \text{ for some natural number } n \geq 1\}$. Clearly, $\emptyset \notin \mathcal{U}$ (and $\mathcal{U}$ is non-empty), so $\mathcal{U}$ is consistent. We have seen that $\mathcal{U}$ is closed under supersets. Finally, $\mathcal{U}$ is easily seen to be closed under finite intersections: let $X_1, X_2 \in \mathcal{U}$. Then, there is an $n \geq 1$ and $m \geq 1$ such that $(-\frac{1}{n}, \frac{1}{n}) \subseteq X_1$ and $(-\frac{1}{m}, \frac{1}{m}) \subseteq X_2$. Either $n \leq m$ or $m > n$. If $n = m$, we are done since in this case, $(-\frac{1}{n}, \frac{1}{n}) = (-\frac{1}{m}, \frac{1}{m}) \subseteq X_1 \cap X_2$ and so $X_1 \cap X_2 \in \mathcal{U}$. This leaves the cases $n > m$ and $m > n$. Suppose that $n > m$. Then $(-\frac{1}{n}, \frac{1}{n}) \subseteq (-\frac{1}{m}, \frac{1}{m})$. Hence, $(-\frac{1}{n}, \frac{1}{n}) \subseteq X_1 \cap X_2$ and so $X_1 \cap X_2 \in \mathcal{U}$. The case in which $m > n$ is similar. So $\mathcal{U}$ is a consistent filter.

Now, $\cap \mathcal{U} = \emptyset$ and, as noted above, $\emptyset \notin \mathcal{U}$; therefore, $\mathcal{U}$ is not augmented. To see that $\cap \mathcal{U} = \emptyset$, note that for each $x \in (-1, 1)$, there is a large enough n such that $x \notin (-\frac{1}{n}, \frac{1}{n})$ (this is a standard fact about real numbers). This shows that $\cap_{n \geq 1}(-\frac{1}{n}, \frac{1}{n}) = \emptyset$. Thus, since $\cap \mathcal{U} \subseteq \cap_{n \geq 1}(-\frac{1}{n}, \frac{1}{n})$, we have $\cap \mathcal{U} = \emptyset$. □

Note that it is crucial that W is infinite in the above example. In fact, as is well known, the situation is much better when W is finite. This is demonstrated by the following Lemma and Corollary (the proofs are left to the reader).

Exercise 1.2 Prove the following Lemma and Corollary:

Lemma 1.2 *If $\mathcal{U}$ is closed under binary intersections (i.e., if $X, Y \in \mathcal{U}$, then $X \cap Y \in \mathcal{U}$), then $\mathcal{U}$ is closed under finite intersections.*

Corollary 1.1 *If W is finite and $\mathcal{U}$ is a filter over W, then $\mathcal{U}$ is augmented.*

I conclude this section with some additional notation. Suppose that W is a non-empty set and $\mathcal{U} \subseteq \wp(W)$ any collection of sets.

- Let $\mathcal{U}^{mon}$ be the smallest collection of subsets of W that contains $\mathcal{U}$ and is closed under supersets.
- Let $\mathcal{U}^{aug}$ be the smallest augmented collection of sets containing $\mathcal{U}$. That is, $\mathcal{U}^{aug} = (\mathcal{U} \cup \{\cap \mathcal{U}\})^{mon}$.

1.2 Language and Semantics

Definition 1.6 (*The Basic Modal Language*) Suppose that $\mathsf{At} = \{p, q, r, \ldots\}$ is a (finite or countable) set of sentence letters, or atomic propositions. The set of well-formed formulas generated from At, denoted $\mathcal{L}(\mathsf{At})$, is the smallest set of formulas generated by the following grammar:

1.2 Language and Semantics

$$p \mid \neg\varphi \mid (\varphi \wedge \psi) \mid \Box\varphi \mid \Diamond\varphi$$

where $p \in \mathsf{At}$.

Additional propositional connectives (e.g., $\vee, \rightarrow, \leftrightarrow$) are defined as usual. It will be convenient to introduce special formulas '$\top$' and '$\bot$', meaning 'true' and 'false', respectively. Typically, $\bot$ is defined to be $p \wedge \neg p$ (where $p \in \mathsf{At}$) and $\top$ is $\neg\bot$.[3] Examples of modal formulas include[4]: $\Box\bot$, $\Box\Diamond\top$, $p \rightarrow \Box(q \wedge r)$, and $\Box(p \rightarrow (q \vee \Diamond r))$. To simplify the notation, I write $\mathcal{L}$ for $\mathcal{L}(\mathsf{At})$ when the set of atomic propositions At is understood.

Remark 1.7 (*Modal Operators*) According to Definition 1.6, $\mathcal{L}$ contains two *unary* modal operators. In this text, I will discuss languages that contain more than two unary modalities and languages that contain modalities of other arities (e.g., the binary modality in Sect. 1.4.3). Furthermore, it is often convenient to *define* $\Diamond\varphi$ as $\neg\Box\neg\varphi$ (cf. Lemma 2.4).

One language, many readings. There are many possible readings for the modal operators '$\Box$' and '$\Diamond$'. Here are some samples:

- **Alethic Reading**: $\Box\varphi$ means 'φ is necessary' and $\Diamond\varphi$ means 'φ is possible'.
- **Deontic Reading**: $\Box\varphi$ means 'φ is obligatory' and $\Diamond\varphi$ means 'φ is permitted'. In this literature, 'O' typically is used instead of '$\Box$' and 'P' instead of '$\Diamond$'.
- **Epistemic Reading**: $\Box\varphi$ means 'φ is known' and $\Diamond\varphi$ means 'φ is consistent with the knower's current information'. In this literature, 'K' typically is used instead of '$\Box$' and 'L' instead of '$\Diamond$'.
- **Temporal Reading**: $\Box\varphi$ means 'φ will always be true' and $\Diamond\varphi$ means 'φ will be true at some point in the future'. In this literature, 'G' typically is used instead of '$\Box$' and 'F' instead of '$\Diamond$'.

I conclude this brief introduction to the basic modal language with the standard definition of a substitution between formulas.

Definition 1.8 (*Substitution*) A **substitution** σ is a function from atomic propositions to well-formed formulas: $\sigma : \mathsf{At} \rightarrow \mathcal{L}(\mathsf{At})$. A substitution σ is extended to a function on all formulas, denoted $\overline{\sigma} : \mathcal{L}(\mathsf{At}) \rightarrow \mathcal{L}(\mathsf{At})$, by recursion on the structure of the formulas:

1. $\overline{\sigma}(p) = \sigma(p)$
2. $\overline{\sigma}(\varphi \wedge \psi) = \overline{\sigma}(\varphi) \wedge \overline{\sigma}(\psi)$
3. $\overline{\sigma}(\Box\varphi) = \Box\overline{\sigma}(\varphi)$
4. $\overline{\sigma}(\Diamond\varphi) = \Diamond\overline{\sigma}(\varphi)$

For simplicity, I will often identify σ and $\overline{\sigma}$ and write φ^σ for $\sigma(\varphi)$.

---

[3] If the set of atomic propositions is empty, then add $\bot$ and $\top$ to the language.
[4] To simplify the presentation, I will typically drop the outermost parentheses.

For example, if $\sigma(p) = \Box\Diamond(p \wedge q)$ and $\sigma(q) = p \wedge \Box q$, then

$$(\Box(p \wedge q) \to \Box p)^\sigma = \Box((\Box\Diamond(p \wedge q)) \wedge (p \wedge \Box q)) \to \Box(\Box\Diamond(p \wedge q)).$$

Exercise 1.3 1. Suppose that $\sigma(p) = \Box q$ and $\sigma(q) = (p \to \Box q)$. Find $(\Box(p \to q) \to (\Box p \to \Box q))^\sigma$.
2. Show that $\varphi^\sigma = \varphi$ iff $\sigma(p) = p$ for all atomic propositions p occurring in φ.
3. Suppose that $((\varphi)^\sigma)^\sigma = \varphi$, but $(\varphi)^\sigma \neq \varphi$. Show that $\varphi \in \mathsf{At}$.

1.2.1 Neighborhood Frames and Models

The definition of a neighborhood model is very simple: Each state from W is associated with a subset space over W.

Definition 1.9 (*Neighborhood Frame*) Let W be a non-empty set. A function $N : W \to \wp(\wp(W))$ is called a **neighborhood function**. A pair $\langle W, N \rangle$ is a called a **neighborhood frame** if W is a non-empty set and N is a neighborhood function.

Remark 1.10 (*Neighborhood Relation*) It is sometimes convenient to treat a neighborhood function $N : W \to \wp(\wp(W))$ as a relation. More precisely, every neighborhood function N corresponds to a relation $R_N \subseteq W \times \wp(W)$ such that for any $w \in W$, $X \in \wp(W)$, $w\ R_N\ X$ iff $X \in N(w)$.

A neighborhood frame $\langle W, N \rangle$ is said to be a filter provided that for each $w \in W$, $N(w)$ is a filter. It is similar for the other properties discussed in Sect. 1.1. Admittedly, this is an abuse of notation since a filter is a property of collections of sets rather than of a neighborhood function. However, I trust that this will not cause any confusion.

Definition 1.11 (*Neighborhood Model*) Suppose that $\mathcal{F} = \langle W, N \rangle$ is a neighborhood frame. A **model** based on $\mathcal{F}$ is a tuple $\langle W, N, V \rangle$, where $V : \mathsf{At} \to \wp(W)$ is a valuation function (assigning a set of states to each atomic proposition).

Definition 1.12 (*Truth*) Suppose that $\mathcal{M} = \langle W, N, V \rangle$ is a neighborhood model and that $w \in W$. Truth of formulas $\varphi \in \mathcal{L}(\mathsf{At})$ at w is defined by recursion on the structure of φ:

1. $\mathcal{M}, w \models p$ iff $w \in V(p)$ $\quad (p \in \mathsf{At})$.
2. $\mathcal{M}, w \models \neg\varphi$ iff $\mathcal{M}, w \not\models \varphi$.
3. $\mathcal{M}, w \models (\varphi \wedge \psi)$ iff $\mathcal{M}, w \models \varphi$ and $\mathcal{M}, w \models \psi$.
4. $\mathcal{M}, w \models \Box\varphi$ iff $[\![\varphi]\!]_\mathcal{M} \in N(w)$.
5. $\mathcal{M}, w \models \Diamond\varphi$ iff $W - [\![\varphi]\!]_\mathcal{M} \notin N(w)$.

where $[\![\varphi]\!]_\mathcal{M}$ is the **truth set** of φ. That is, $[\![\varphi]\!]_\mathcal{M} = \{w \mid \mathcal{M}, w \models \varphi\}$. A set of formulas $\Gamma \subseteq \mathcal{L}$ is **satisfiable** if there is some model $\mathcal{M} = \langle W, N, V \rangle$ and world $w \in W$ such that $\mathcal{M}, w \models \varphi$ for all $\varphi \in \Gamma$. A formula $\varphi \in \mathcal{L}$ is satisfiable when $\{\varphi\}$ is satisfiable.

1.2 Language and Semantics

Definition 1.13 (*Validity*) Suppose that $\mathcal{M} = \langle W, N, V \rangle$ is a neighborhood model. A formula $\varphi \in \mathcal{L}$ is **valid on** $\mathcal{M}$, denoted $\mathcal{M} \models \varphi$, when $\mathcal{M}, w \models \varphi$ for all $w \in W$. Suppose that $\mathcal{F} = \langle W, N \rangle$ is a neighborhood frame. For each $w \in W$, a formula φ is **valid at** w **in** $\mathcal{F}$, denoted $\mathcal{F}, w \models \varphi$, provided that $\mathcal{M}, w \models \varphi$ for all models $\mathcal{M}$ based on $\mathcal{F}$ (i.e., $\mathcal{M} = \langle \mathcal{F}, V \rangle$). A formula $\varphi \in \mathcal{L}$ is **valid on** $\mathcal{F}$, denoted $\mathcal{F} \models \varphi$, provided that $\mathcal{F}, w \models \varphi$ for all $w \in W$. Suppose that C is a class of frames. A formula $\varphi \in \mathcal{L}$ is valid on C, denoted $\models_\mathsf{C} \varphi$, provided that $\mathcal{F} \models \varphi$ for all $\mathcal{F} \in \mathsf{C}$.

The definition of truth for the modal operators (items 4 and 5 in Definition 1.12) was chosen to ensure that $\Box$ and $\Diamond$ are *duals* (cf. Lemma 2.4). The basic idea is that the neighborhood function N lists, for each state, the propositions considered "necessary." Then, $\Box \varphi$ is true at a state when the truth set of φ is a member of that list at the state. Furthermore, φ is "possible" if the proposition expressed by $\neg \varphi$ is not a member of the list at the state. Other options for the definition of truth of the modal operators have been considered in the literature. I will consider some of these below. Let us start by getting a feel for the above definition of truth.

We first need some notation. Each neighborhood function $N : W \to \wp(\wp(W))$ is associated with a function $m_N : \wp(W) \to \wp(W)$ as follows:

$$\text{for } X \subseteq W, \; m_N(X) = \{w \mid X \in N(w)\}.$$

Intuitively, $m_N(X)$ is the set of states in which X is necessary. Let $\mathcal{M} = \langle W, N, V \rangle$ be a neighborhood model. Then,

$$[\![p]\!]_\mathcal{M} = V(p) \text{ for } p \in \mathsf{At}.$$
$$[\![\neg \varphi]\!]_\mathcal{M} = W - [\![\varphi]\!]_\mathcal{M}.$$
$$[\![(\varphi \land \psi)]\!]_\mathcal{M} = [\![\varphi]\!]_\mathcal{M} \cap [\![\psi]\!]_\mathcal{M}.$$
$$[\![\Box \varphi]\!]_\mathcal{M} = m_N([\![\varphi]\!]_\mathcal{M}).$$
$$[\![\Diamond \varphi]\!]_\mathcal{M} = W - m_N(W - [\![\varphi]\!]_\mathcal{M}).$$

It is an easy application of the definition of truth to verify the above equations. I leave this proof as an exercise for the reader. The following example illustrates the definitions of models and truth of modal formulas:

Example 1.14 (*Detailed Example of a Neighborhood Model*) Suppose that $W = \{w, s, v\}$, and define a neighborhood function $N : W \to \wp(\wp(W))$ as follows:

- $N(w) = \{\{s\}, \{v\}, \{w, v\}\}$.
- $N(s) = \{\{w, v\}, \{w, s\}, \{w\}\}$.
- $N(v) = \{\{w\}, \{s, v\}, \emptyset\}$.

Define a valuation function $V : \{p, q\} \to \wp(W)$ by $V(p) = \{w, s\}$ and $V(q) = \{s, v\}$. Then, $\mathcal{M} = \langle W, N, V \rangle$ is a neighborhood model. This model can be depicted as follows:

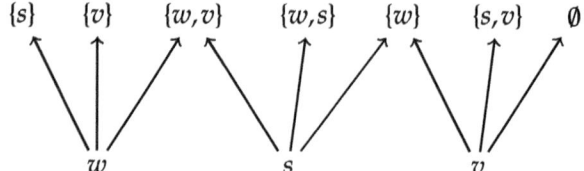

We can now calculate the truth (using Definition 1.12) for various modal formulas.

1. Since $[\![p]\!]_\mathcal{M} = V(p) = \{w, s\} \in N(s)$, we have $\mathcal{M}, s \models \Box p$.
2. Since $[\![\neg p]\!]_\mathcal{M} = \{v\} \notin N(s)$, we have $\mathcal{M}, s \not\models \Diamond p$.
3. Since $[\![\Diamond p]\!]_\mathcal{M} = \{s, v\} \in N(v)$, we have $\mathcal{M}, v \models \Box\Diamond p$.
4. Since $[\![\Box p]\!]_\mathcal{M} = \{s\} \in N(w)$, we have $\mathcal{M}, w \models \Box\Box p$.
5. Since $[\![\Box\Box p]\!]_\mathcal{M} = \{w\} \in N(s) \cap N(v)$, we have $\mathcal{M}, s \models \Box\Box\Box p$ and $\mathcal{M}, v \models \Box\Box\Box p$.
6. Since $[\![\Box\Box\Box p]\!]_\mathcal{M} = \{w, v\} \in N(w) \cap N(s)$, we have $\mathcal{M}, w \models \Box\Box\Box\Box p$ and $\mathcal{M}, s \models \Box\Box\Box\Box p$.
7. Finally, since $[\![\bot]\!]_\mathcal{M} = \emptyset \in N(v)$, we have $\mathcal{M}, v \models \Box\bot$.

Note that $\mathcal{M}, w \models \Box(p \wedge q)$, but $\mathcal{M}, w \not\models \Box p$. This is the first difference between neighborhood models and relational models: If we fix the valuations of p and q as in the above example, it is not possible to define a relational structure such that $\Box(p \wedge q)$ is true at w but $\Box p$ is false at w. Let us see why. (Consult Appendix A for an introduction to relational semantics for modal logic.) Assuming that $\Box p$ is false at w forces w to have an accessible world in which p is false. But, according to the above valuation, there is only one such world in which p is false—namely, v. However, if v is accessible from w, then $\Box(p \wedge q)$ will no longer be true at w (since, if p is false at v, then so is $p \wedge q$).

Exercise 1.4 1. Consider the neighborhood model $\mathcal{M} = \langle W, N, V \rangle$ where $W = \{w, v, x\}$; $N(w) = \{\{w, v\}, \{v, x\}\}$, $N(v) = \emptyset$, and $N(x) = \{\{w, v\}, \{v, x\}, \{v\}\}$; and $V(p) = \{w, v\}$ and $V(q) = \{v, x\}$. Find the truth sets of the following modal formulas: $\Box p, \Box q, \Box(p \wedge q), \Box p \wedge \Box q$, and $\Diamond p$.
2. For each of the following formulas, find a neighborhood models that falsifies the formula (i.e., find a model that contains a state in which the formula is false): $\Box p \to \Diamond p, \Diamond p \to \Diamond(p \vee q)$, and $\Box\top$
3. Prove that $\Box(\varphi \wedge \psi) \to \Box\varphi$ is valid on a neighborhood frame iff $\Diamond\varphi \to \Diamond(\varphi \vee \psi)$ is valid on the frame.

Exercise 1.5 Suppose that $\mathcal{M} = \langle W, N, V \rangle$ and $\mathcal{M}' = \langle W, N, V' \rangle$ are two neighborhood models based on the same neighborhood frame $\langle W, N \rangle$. Suppose that $p_1, p_2, \ldots, p_k$ are the atomic propositions in φ and $\psi_1, \ldots, \psi_k$ are formulas such that for all $i = 1, \ldots, k$, $V(p_i) = [\![\psi_i]\!]_{\mathcal{M}'}$. If σ is a substitution such that $\sigma(p_i) = \psi_i$, then $[\![\psi]\!]_\mathcal{M} = [\![\psi^\sigma]\!]_{\mathcal{M}'}$.

1.2.2 Additional Modal Operators

Alternative definitions of truth for the basic modal operators can be found in the literature. In particular, David Lewis (1973) introduced a variety of modal operators interpreted on neighborhood models (including the ones defined below) in his seminal book *Counterfactuals* (see Sect. 1.4.3 for a discussion). In order to compare these different definitions, I extend the basic modal language with the following modalities: $[\,)$, $\langle\,]$, $\langle\,\rangle$, and $[\,]$. Let $\mathcal{M} = \langle W, N, V \rangle$ be a neighborhood structure. Truth at a state $w \in W$ for these modalities is given below.

- $\mathcal{M}, w \models \langle\,]\varphi$ iff there is an $X \in N(w)$ such that for all $v \in X$, $\mathcal{M}, v \models \varphi$.
- $\mathcal{M}, w \models [\,)\varphi$ iff for all $X \in N(w)$, there is a $v \in X$ such that $\mathcal{M}, v \models \varphi$.
- $\mathcal{M}, w \models \langle\,\rangle\varphi$ iff there is an $X \in N(w)$ such that there is a $v \in X$, such that $\mathcal{M}, v \models \varphi$.
- $\mathcal{M}, w \models [\,]\varphi$ iff for all $X \in N(w)$, for all $v \in X$, $\mathcal{M}, v \models \varphi$.

The first observation is that there really are only two modalities.

Observation 1.15 *The following formulas are valid on all neighborhood models.*

- $\langle\,]\varphi \leftrightarrow \neg[\,)\neg\varphi$.
- $[\,]\varphi \leftrightarrow \neg\langle\,\rangle\neg\varphi$.

Exercise 1.6 Prove the above observation.

The modalities $\langle\,]$ and $[\,)$ will play an important role throughout this book. Let $\mathcal{L}^{mon}(\mathsf{At})$ be the modal language generated by the following grammar:

$$p \mid \neg\varphi \mid (\varphi \wedge \psi) \mid \langle\,]\varphi \mid [\,)\varphi$$

where $p \in \mathsf{At}$.

Lemma 1.3 *Let $\mathcal{M} = \langle W, N, V \rangle$ be a neighborhood model. Then, for each $w \in W$,*

1. *if $\mathcal{M}, w \models \Box\varphi$ then $\mathcal{M}, w \models \langle\,]\varphi$; and*
2. *if $\mathcal{M}, w \models [\,)\varphi$ then $\mathcal{M}, w \models \Diamond\varphi$.*

Proof Let $\mathcal{M} = \langle W, N, V \rangle$ be a neighborhood model and $w \in W$. Suppose that $\mathcal{M}, w \models \Box\varphi$. Then $[\![\varphi]\!]_{\mathcal{M}} \in N(w)$. Then, clearly, there is an $X \in N(w)$ such that for each $v \in X$, $\mathcal{M}, v \models \varphi$ (let $X = [\![\varphi]\!]_{\mathcal{M}}$). The proof of the second statement is analogous, and so will be left to the reader. □

The converses of both statements in the above Lemma are false. For instance, note that in Example 1.14, $\mathcal{M}, w \models \langle\,]p$ (this follows since $\{s\} \in N(w)$ and $\{s\} \subseteq [\![p]\!]_{\mathcal{M}} = \{w, s\}$). Thus, $\langle\,]\varphi \to \Box\varphi$ is not valid on neighborhood models. This shows that the two definitions for a modal operator are not, in general, equivalent. However, they are equivalent when restricting attention to monotonic neighborhood frames (you are asked to prove this in Exercise 1.7). As we will see in Chap. 2, there are theoretical reasons to prefer working in languages with $\langle\,]$-modalities (cf. Areces and Figueira, 2009).

Exercise 1.7 1. Prove that if $\varphi \to \psi$ is valid, then so is $\langle\,]\varphi \to \langle\,]\psi$.
2. Suppose that $\mathcal{F} = \langle W, N\rangle$ is a neighborhood frame. Prove that $\Box\varphi \leftrightarrow \langle\,]\varphi$ is valid on $\mathcal{F}$ iff $\mathcal{F}$ is monotonic.
3. Are there analogous results for the $\langle\,\rangle$ modality?

1.2.3 Reasoning About Subset Spaces

Larry Moss and Rohit Parikh devised a simple modal language with two modalities (K and $\Diamond$) for reasoning about subset spaces (Moss and Parikh, 1992). Moss and Parikh had an epistemic interpretation in mind. The idea is that, given a subset space $\langle W, \mathcal{O}\rangle$, the set $\mathcal{O}$ represents the set of all observations about the states W available to an agent. Formulas of their language are interpreted at pairs (w, U), where w is a state and U is set with $w \in U$ representing the agent's current observation. Thus, it is assumed that all observations are *reliable*, in the sense that the actual world is always an element of the agent's current observation. The intended interpretation of $K\varphi$ is that "φ is known, given the agent's current observation", and the intended interpretation of $\Diamond\varphi$ is that "after some effort (such as an additional measurement), φ becomes true". For example, the formula

$$\varphi \to \Diamond K\varphi$$

means that if φ is true, then after some "effort", $K\varphi$ is true. In other words, the formula says that if φ is true, then φ can be known with some effort. What exactly is meant by "effort" depends on the application. I will now give the formal definitions of the syntax and semantics from (Moss and Parikh, 1992).

Definition 1.16 (*Subset Space Modal Language*) Suppose that At is a finite or countable set of atomic propositions. Let $\mathcal{L}^{K\Diamond}(\mathsf{At})$ be the smallest set of formulas generated by the following grammar:

$$p \mid \neg\varphi \mid (\varphi \wedge \psi) \mid K\varphi \mid \Diamond\varphi$$

where $p \in \mathsf{At}$. The other Boolean connectives are defined as usual. In addition, the duals of the modal operators are defined as follows: $L\varphi$ is $\neg K\neg\varphi$ and $\Box\varphi$ is $\neg\Diamond\neg\varphi$.

Definition 1.17 (*Subset Space Model*) A **subset space model** is a tuple $\langle W, \mathcal{O}, V\rangle$, where W is a non-empty set, $\mathcal{O} \subseteq \wp(W)$ and $V : \mathsf{At} \to \wp(W)$ is a valuation function. Let $W \dot\times \mathcal{O} = \{(w, U) \mid w \in W,\ U \in \mathcal{O},\ \text{and}\ w \in U\}$, elements of which are called **neighborhood situations**.

Definition 1.18 (*Truth in a Subset Space Model*) Let $\mathcal{M} = \langle W, \mathcal{O}, V\rangle$ be a subset space model and $(w, U) \in W \dot\times \mathcal{O}$ a neighborhood situation. Truth of formulas $\varphi \in \mathcal{L}^{K\Diamond}(\mathsf{At})$ at (w, U) is defined by induction on the structure of φ.

1. $\mathcal{M}, w, U \models p$ iff $w \in V(p)$ where $p \in \mathsf{At}$.

2. $\mathcal{M}, w, U \models \neg\varphi$ iff $\mathcal{M}, w, U \not\models \varphi$.
3. $\mathcal{M}, w, U \models \varphi \wedge \psi$ iff $\mathcal{M}, w, U \models \varphi$ and $\mathcal{M}, w, U \models \psi$.
4. $\mathcal{M}, w, U \models K\varphi$ iff for all $v \in U$, $\mathcal{M}, v, U \models \varphi$.
5. $\mathcal{M}, w, U \models \Diamond\varphi$ iff there is a $(w, V) \in W \dot\times \mathcal{O}$ such that $V \subseteq U$ and $\mathcal{M}, w, V \models \varphi$.

The usual logical notions of validity and satisfiability are defined in the standard way.

Note that the interpretation of the atomic formulas at a neighborhood situation (w, U) does not depend on the second component U. This means that, in any subset space model, if φ does not contain any modalities, then $\varphi \leftrightarrow \Box\varphi$ is valid. Thus, effort will not change the ground facts about the world—it can only change knowledge of these facts. I conclude by highlighting some of the properties of subset spaces that can be expressed in the above bi-modal language.

Observation 1.19 *The axiom scheme $K\Box\varphi \rightarrow \Box K\varphi$ is valid on any subset space model.*

Proof Suppose that $\mathcal{M} = \langle W, \mathcal{O}, V \rangle$ is a subset space model and (w, U) is a neighborhood situation with $\mathcal{M}, w, U \models K\Box\varphi$. Then, for all $v \in U$, $\mathcal{M}, v, U \models \Box\varphi$. This means that for all $v \in U$ and all $V \in \mathcal{O}$, if $v \in V \subseteq U$, then $\mathcal{M}, v, V \models \varphi$. Let $(w, U') \in W \dot\times \mathcal{O}$ be any neighborhood situation in which $U' \subseteq U$. We must show that $\mathcal{M}, w, U' \models K\varphi$. Let $y \in U'$ be any state in U'. Since $y \in U' \subseteq U$, by the assumption, $\mathcal{M}, y, U \models \Box\varphi$. This means that, since $y \in U' \subseteq U$, $\mathcal{M}, y, U' \models \varphi$. Hence, $\mathcal{M}, w, U' \models K\varphi$, and $\mathcal{M}, w, U \models \Box K\varphi$. Thus, $K\Box\varphi \rightarrow \Box K\varphi$ is valid. □

It is not difficult to see that the K modality validates the so-called **S5** axioms $K(\varphi \rightarrow \psi) \rightarrow (K\varphi \rightarrow K\psi)$, $K\varphi \rightarrow \varphi$, $K\varphi \rightarrow KK\varphi$ and $\neg K\varphi \rightarrow K\neg K\varphi$; and that the $\Box$ modality validates the so-called **S4** axioms $\Box(\varphi \rightarrow \psi) \rightarrow (\Box\varphi \rightarrow \Box\psi)$, $\Box\varphi \rightarrow \varphi$, and $\Box\varphi \rightarrow \Box\Box\varphi$ (cf. Sect. A.2). Moss and Parikh (1992) proved that these axioms, together with the axiom scheme from Observation 1.19 ($K\Box\varphi \rightarrow \Box K\varphi$) and rules of necessitation for each modality, completely axiomatize the class of all subset space models. Additional properties of subset spaces can be expressed in the language.

Observation 1.20 *The axiom scheme $\Box\Diamond\varphi \rightarrow \Diamond\Box\varphi$ is valid on any subset space that is closed under arbitrary intersections.*

Proof Let $\mathcal{M} = \langle W, \mathcal{O}, V \rangle$ be a subset space in which $\mathcal{O}$ is closed under arbitrary intersections. Suppose that $\mathcal{M}, w, U \models \Box\Diamond\varphi$. Then, for all $w, V \in W \dot\times \mathcal{O}$, if $V \subseteq U$, then $\mathcal{M}, w, V \models \Diamond\varphi$. We must show that $\mathcal{M}, w, U \models \Diamond\Box\varphi$. Let $\mathcal{U}_w = \{U \mid U \in \mathcal{O} \text{ and } w \in U\}$ be the set of elements of $\mathcal{O}$ that contain w. Since $\mathcal{O}$ is closed under arbitrary intersections, we have $\bigcap \mathcal{U}_w \in \mathcal{O}$. Note that $\bigcap \mathcal{U}_w$ does not have any proper subsets in $\mathcal{O}$. Thus, since $\mathcal{M}, w, \bigcap \mathcal{U}_w \models \varphi$ or $\mathcal{M}, w, \bigcap \mathcal{U}_w \models \neg\varphi$, we have $\mathcal{M}, w, \bigcap \mathcal{U}_w \models \Box\varphi \vee \Box\neg\varphi$. By the assumption, since $w \in \bigcap \mathcal{U}_w \subseteq U$, we have $\mathcal{M}, w, \bigcap \mathcal{U}_w \models \Diamond\varphi$. Since $\mathcal{M}, w, \bigcap \mathcal{U}_w \models \Box\varphi \vee \Box\neg\varphi$ and $\mathcal{M}, w, \bigcap \mathcal{U}_w \models \Diamond\varphi$, it must be the case that $\mathcal{M}, w, \bigcap \mathcal{U}_w \models \varphi$ (this follows since $\bigcap \mathcal{U}_w$ is a subset of itself).

Thus, since there are no proper subsets of $\bigcap \mathcal{U}_w$ in $\mathcal{O}$, we have $\mathcal{M}, w, \bigcap \mathcal{U}_w \models \Box \varphi$. This means that $\mathcal{M}, w, U \models \Diamond \Box \varphi$, as desired. Hence, $\Box \Diamond \varphi \to \Diamond \Box \varphi$ is valid when the collection of subsets are closed under arbitrary intersections. □

Interestingly, it can be shown that an infinite number of axiom schemes are needed for a complete axiomatization of subset spaces that are closed under arbitrary intersections (Weiss and Parikh, 2002).

Exercise 1.8 1. Prove that $(\Diamond \varphi \wedge L \Diamond \psi) \to \Diamond [\Diamond \varphi \wedge L \Diamond \psi \wedge K \Diamond L(\varphi \vee \psi)]$ is valid on any subset space that is closed under binary unions.
2. Say that a subset space is *directed* provided that for all $w \in W$ and $U, V \in \mathcal{O}$ with $w \in U \cap V$, there is a $X \in \mathcal{O}$ such that $w \in X \subseteq U \cap V$. Prove that $\Diamond \Box \varphi \to \Box \Diamond \varphi$ is valid on all subset spaces that are directed.

A number of papers have focused on finding axiomatizations of different classes of subsets spaces and increasing the expressive power of the subset space language $\mathcal{L}^{K \Diamond}$. See Georgatos (1993) for a sound and complete axiomatization of subset spaces that are topologies and Georgatos (1994) for a sound and complete axiomatization for subsets spaces that are complete lattices. The basic subset logic framework introduced in this section has been extended to include temporal operators (Heinemann, 1999, 2000) and hybrid modalities (Heinemann, 2004). In addition, there are subset space logics with public announcement operators and many agents (Wáng and Agotnes, 2013; van Ditmarsch et al., 2015; Bjorndahl, 2016). Consult Moss et al. (2007) for further discussion of logics for reasoning about subset spaces.

1.3 Why Non-normal Modal Logic?

The following formulas and rules of inference are valid on all relational models (see Appendix A for a discussion):

(Dual) $\Box \varphi \leftrightarrow \neg \Diamond \neg \varphi$ (RE) From $\varphi \leftrightarrow \psi$, infer $\Box \varphi \leftrightarrow \Box \psi$
(C) $(\Box \varphi \wedge \Box \psi) \to \Box(\varphi \wedge \psi)$ (RM) From $\varphi \to \psi$, infer $\Box \varphi \to \Box \psi$
(K) $\Box(\varphi \to \psi) \to (\Box \varphi \to \Box \psi)$ (Nec) From φ, infer $\Box \varphi$

For many interpretations, these formulas and rules are relatively uncontroversial. However, in some interpretations of the basic modal language, the validity of one or more of the above formulas and rules of inference can be questioned. Modal logics that do not include one or more of the above formulas and/or rules are called *non-normal modal logics*. (I will define *normal* and *non-normal* modal logics in Sect. 2.3.) I want to stress that the examples discussed in this section are *not* intended to motivate the use of *neighborhood structures* as a semantics for these weak systems of modal logic. Indeed, the most convincing analyses of many of the examples discussed below do not use neighborhood structures.

1.3 Why Non-normal Modal Logic?

Logical Omniscience

Epistemic logicians interpret $\Box \varphi$ as "the agent knows that φ is true" (see Fagin et al. (1995) and Pacuit (2013a) for discussions and references to the relevant literature). Under this interpretation, the above principles each express a significant assumption about the reasoning abilities of the agent under consideration.

Closure under logical implication (RM): Suppose that $\varphi \to \psi$ is valid. Then, if the agent knows that φ, then the agent knows that ψ. This means that the agent knows all the logical consequences of her knowledge.

Closure under known implication (K): If the agent knows that φ implies ψ and the agent knows that φ, then the agent knows that ψ. Note the difference from the inference rule (RM). The axiom K means that the set of formulas that the agent knows is deductively closed (i.e., if φ and $\varphi \to \psi$ is in the set of formulas known by the agent, then so is ψ). This is weaker than what is imposed under the inference rule (RM): If φ is a formula known by the agent and $\varphi \to \psi$ is valid (but not necessarily in the set of formulas known by the agent), then ψ is also known by the agent.

Closure under logical equivalence (RE): If φ and ψ are logically equivalent (so express the same *proposition*[5]), then the agent knows that φ if, and only if, the agent knows that ψ. This means that the agent cannot distinguish between two different formulas that are logically equivalent.

Knowledge of logical validities (Nec): If φ is valid, then the agent knows that φ. This means that the agent knows anything that can be deduced using the given modal system. In particular, this means that the agent knows all propositional tautologies.

Closure under conjunction (C): If the agent knows that φ and the agent knows that ψ, then the agent knows that $\varphi \wedge \psi$. That is, the set of formulas that the agent knows is closed under conjunction. This is also called *agglomeration*.

Each of the above principles identifies different ways in which the agents studied by epistemic logicians are *idealized* reasoners. There are two main reasons why one may want to drop one or more of the above assumptions about the agents' reasoning abilities. The first reason is that the agents under consideration may not, in fact, be perfect reasoners. For instance, humans typically do not recognize *all* the logical consequences of what they currently know.[6] How, exactly, to reason about agents that are not perfect reasoners is known as the *logical omniscience problem*. This is a difficult problem that is not easily solved by simply restricting what the agents know at each possible world using neighborhood models (see, Halpern and Pucella 2011 for an overview of the different approaches to the logical omniscience problem).

---

[5] Here, I understand a proposition as a set of possible worlds. Alternatively (using standard terminology from probability theory), I will say that φ and ψ express the same *event*.

[6] Indeed, Gilbert Harman famously argued that a rational thinker *should not* make all possible deductions because they "clutter the mind" with useless facts (Harman, 1986).

Even if it is assumed that the agents are perfect reasoners, epistemologists have identified arguments that purport to show that *knowledge* may not satisfy all of the above principles. See Holliday (2012, 2014) for an illuminating discussion of this second reason to drop one or more of the above principles and an overview of the various philosophical positions.

Logics of Knowledge and Belief

Logics of knowledge and belief include modalities for knowledge ($[K]$) and belief ($[B]$). The duals are denoted $\langle K \rangle \varphi$ and $\langle B \rangle \varphi$ and are defined as $\neg [K] \neg \varphi$ and $\neg [B] \neg \varphi$, respectively. In an influential paper, Robert Stalnaker (2006) proposed the following axioms for a logic of knowledge and belief. The first group of axioms defines $[K]$ as an **S4**-modality (cf. Sect. A.2):

(K) $[K](\varphi \to \psi) \to ([K]\varphi \to [K]\psi)$
(T) $[K]\varphi \to \varphi$
(4) $[K]\varphi \to [K][K]\varphi$
(Nec) From φ infer $[K]\varphi$

The second group of axioms characterize the relationship between knowledge and beliefs:

(PI) $[B]\varphi \to [K][B]\varphi$
(NI) $\neg[B]\varphi \to [K]\neg[B]\varphi$
(KB) $[K]\varphi \to [B]\varphi$
(D) $[B]\varphi \to \langle B \rangle \varphi$
(SB) $[B]\varphi \to [B][K]\varphi$

The axioms (PI) and (NI) ensure that the agent's beliefs are perfectly introspective: If the agent believes that φ, then she knows that she believes that φ, and if the agent does not believe that φ, then she knows that she does not believe that φ. The (KB) axiom is the natural idea that knowing something implies that it is believed. The (D) axiom guarantees that the agent's beliefs are consistent: If the agent believes that φ, then it is not the case that the agent believes that $\neg \varphi$. Finally, the (SB) axiom is characteristic of the strong form of belief that Stalnaker has in mind: If the agent believes that φ, then she believes that she knows that φ. Consult Stalnaker (2006) for further motivation and discussion of these axioms. The important point here is that it turns out that the belief operator is definable as possible knowledge. More formally, the following formula is derivable (see Sect. 2.3 for a formal definition of derivations) in Stalnaker's logic:

(DefKB) $[B]\varphi \leftrightarrow \langle K \rangle [K]\varphi$,

The derivation of this formula is left to the reader. It can also be shown that Stalnaker's belief operator defined using (**DefKB**) is a normal modal operator—i.e.,

1.3 Why Non-normal Modal Logic?

the belief operator validates all the axioms and rules mentioned in the introduction to this section. Furthermore, notice that by substituting $\langle K \rangle [K] \varphi$ for $[B]\varphi$ in (D), the following formula is derivable:

$$(.2) \quad \langle K \rangle [K] \varphi \to [K] \langle K \rangle \varphi.$$

Stalnaker's proposal is that knowledge is axiomatized by the **S4** axioms and rules together with the (.2) axiom (this logic is called **S4.2**). The other axioms listed above act as bridges principles relating knowledge and belief.[7]

Exercise 1.9 *(This exercise requires knowledge of relational semantics, see Appendix A.)* Consider the modal formula $(\Diamond \Box p \wedge \Diamond \Box q) \to \Diamond \Box (p \wedge q)$.

1. Find a relational model with a state in which the above formula is not true.
2. Find an **S4**-relational model (i.e., a relational model with a reflexive and transitive relation) with a state in which the above formula is false.

This analysis suggests an interesting way to define non-normal modal operators. Fix a (normal) modal logic **L** with modalities $\Box_1, \ldots, \Box_n$. Define a new modality as a sequence of modalities (or their duals) from **L**. In many cases, this will result in a non-normal modal operator. Indeed, there is a sense in which every non-normal modal logic is generated this way (cf. Gasquet and Herzig (1996), Kracht and Wolter (1999) and Sect. 2.6.2).

Logics of High Probability

Suppose that the interpretation of $\Box \varphi$ is "φ is assigned 'high' probability", where "high" probability means above a certain threshold $r \in [0, 1]$. Under this interpretation, it is not hard to find a counterexample to the axiom scheme (C). Suppose that p and q represent *independent events* (i.e., the probability of $p \wedge q$ is equal to the probability of p times the probability of q), and suppose that the threshold is $r = \frac{2}{3}$. If both p and q are assigned probability $\frac{3}{4}$, then $\Box p$ and $\Box q$ are both true (i.e., the probabilities of p and q are greater than the threshold). However, the probability of $p \wedge q$ is $\frac{3}{4} \times \frac{3}{4} = \frac{9}{16}$, which is less than the threshold. Thus, $\Box(p \wedge q)$ is not true; and, therefore, $(\Box p \wedge \Box q) \to \Box(p \wedge q)$ is not true under this interpretation. Note that this argument does not work if we take the threshold to be 1.[8]

Deontic Logic

Standard deontic logic interprets $\Box \varphi$ as "it ought to be the case that φ". The following example shows that under this interpretation, the monotonicity rule (RM) should *not* be valid. The example is known as the "Paradox of Gentle Murder" introduced by J. Forrester (1984).[9] Consider the following three statements:

---

[7]Consult Baltag et al. (2015, 2013) for an interesting topological interpretation of this modal logic. Another interesting line of inquiry is to study subsystems of Stalnaker's logic (Klein et al., 2015).
[8]What happens if $r < \frac{1}{2}$?
[9]See, also, Goble (1991, 2004) for discussions.

1. Jones murders Smith.
2. Jones ought not to murder Smith.
3. If Jones murders Smith, then Jones ought to murder Smith gently.

Intuitively, these sentences appear to be consistent. However, 1 and 3 together imply that

4. Jones ought to murder Smith gently.

Also we accept the following conditional:

5. If Jones murders Smith gently, then Jones murders Smith.

Of course, this *not* a logical validity but, rather, a fact about the world we live in. Now, if we assume that the monotonicity rule is valid, then statement 5 entails

6. If Jones ought to murder Smith gently, then Jones ought to murder Smith.

And so, statements 4 and 6 together imply

7. Jones ought to murder Smith.

But this contradicts statement 2. The above argument suggests that classical deontic logic should *not* validate the monotonicity rule.

Logics of Ability

There is a long history developing modal logics to reason about individual *abilities* (see Horty 2001; Pörn 1977; Carr 1979; Governatori and Rotolo 2005, and references therein). Without going into the intricacies of what it means for an individual to exercise an "ability", let $Abl_i \varphi$ mean that "the agent i has the ability to do something that makes φ true".

If we want Abl_i to be a modal operator interpreted on a relational structure, we first have to decide whether it is a 'box' or a 'diamond' operator. The following two examples demonstrate that neither interpretation is appropriate for the ability operator Abl_i. Consider axiom (C) $(\Box\varphi \wedge \Box\psi) \rightarrow \Box(\varphi \wedge \psi)$ and its dual $\Diamond(\varphi \vee \psi) \rightarrow \Diamond\varphi \vee \Diamond\psi$. Both are valid on all relational structures. The following examples show that reasoning about abilities does not fit either of these patterns.

Example 1.21 (*Counterexample for $Abl_i(\varphi \vee \psi) \rightarrow (Abl_i\varphi \vee Abl_i\psi)$*) Suppose that Ann is drawing cards from a normal deck of 52 cards. Let R be the proposition "Ann draws a red card" and B the proposition "Ann draws a black card". Now, since we are assuming that Ann has the ability to choose a card, $Abl_A(R \vee B)$ is true: Ann can pick a card that, as a matter of fact, is either red or black. That is, she has the ability to take some action (pick a card) that makes $R \vee B$ true. However, assuming that Ann is a normal card player and is using a standard deck of cards, there is no way that Ann can select a card that *guarantees* that the card will be red (black). So, she has neither the ability to choose a red card ($\neg Abl_A R$) nor the ability to choose a black card ($\neg Abl_A B$). Thus, $Abl_A(R \vee B) \rightarrow (Abl_A R \vee Abl_A B)$ is not true in this situation.

1.3 Why Non-normal Modal Logic?

Example 1.22 (*Counterexample for* $(Abl_i\varphi \wedge Abl_i\psi) \rightarrow Abl_i(\varphi \wedge \psi)$) Consider the following game between Ann (*A*) and Bob (*B*)

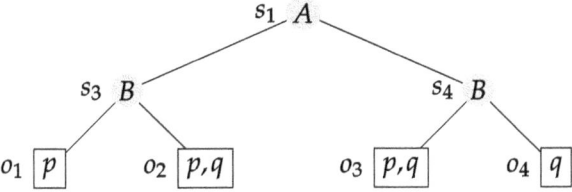

Here, Ann has the first move, while Bob has the next. The game model also indicates which propositional variables (p and q) are true at the outcome nodes (labeled o_1, o_2, o_3 and o_4).

We say that a player has the ability to *force* a set of outcome states X if that player has a *strategy* that guarantees that the game will end in one of the states in X. For example, in the above game, Ann has the ability to force the set $X_1 = \{o_1, o_2\}$. This follows because Ann has a strategy (move left) such that no matter what action Bob chooses, the outcome of the game will be a state in X_1. Similarly, Ann can also force the set $X_2 = \{o_3, o_4\}$. However, Ann cannot force the set $X_3 = \{o_2, o_3\}$ since Bob has the freedom to select either o_1 or o_4 depending on Ann's choice. Note that p is true at states $\{o_1, o_2, o_3\}$ and q is true at $\{o_2, o_3, o_4\}$. Thus, Ann has the ability to force p and the ability to force q ($Abl_A p \wedge Abl_A q$), but she does not have the ability to force both p and q ($\neg Abl_A(p \wedge q)$).

Motivated by the game-theoretic example used in Example 1.22, we can see that (**Dual**) expresses an non-trivial fact about games. Often, $\Diamond\varphi$ is simply assumed to be *defined* as $\neg\Box\neg\varphi$. The game-theoretical interpretation introduced in Example 1.22 suggests the following interpretation of $\Diamond\varphi$: "*Bob* has a strategy to ensure that φ is true". A natural assumption about a game is consistency: it cannot be the case that Ann can force φ to be true and Bob can force $\neg\varphi$ to be true. Thus, $\neg(\Box\varphi \wedge \Diamond\neg\varphi)$ expresses that the game is consistent. Using propositional reasoning, this formula is equivalent to $\Box\varphi \rightarrow \neg\Diamond\neg\varphi$. The converse expresses a stronger game-theoretic assumption. Rewriting $\neg\Diamond\neg\varphi \rightarrow \Box\varphi$ as $\Diamond\neg\varphi \vee \Box\varphi$, this formula says that "either Bob has a strategy to force $\neg\varphi$ to be true or Ann has a strategy to force φ to be true." If we think of the formula φ as stating that Ann has won the game, then this formula expresses that the game is *determined* (either Ann or Bob has a winning strategy). Thus, $\Box\varphi \leftrightarrow \neg\Diamond\neg\varphi$ is true for all consistent, two-person determined games.

The Logic of Classical Deduction

Naumov (2006) introduced a modal logic of classical deduction. Suppose that At is a set of atomic propositions and that $\mathcal{L}_0(\text{At}) \subseteq \mathcal{L}(\text{At})$ is the set of propositional formulas generated from At. Let $\vdash_0$ denote the propositional consequence relation. Fix a set of propositional formulas $\Sigma \subseteq \mathcal{L}_0(\text{At})$. An **interpretation** is a function $(\cdot)^* : \text{At} \rightarrow \wp(\Sigma)$ assigning a set of formulas from Σ to each atomic proposition. An interpretation is extended to all modal formulas in $\mathcal{L}(\text{At})$ as follows:

- $(\varphi \wedge \psi)^* = (\varphi)^* \cap (\psi)^*$.
- $(\neg \varphi)^* = \Sigma - (\varphi)^*$.
- $(\Box \varphi)^* = \{\alpha \in \Sigma \mid (\varphi)^* \vdash_0 \alpha\}$.

So, $\Box \varphi$ denotes the set of propositional consequences (in the universe Σ) of the interpretation of φ. It is not hard to see that under this interpretation, the axiom (C) is not valid. Suppose that Σ is the set of propositional formulas generated from the set $\{p, q\}$. Fix an interpretation with $(p)^* = \{p\}$ and $(q)^* = \{q\}$. Then,

$$(p \vee q) \in (\Box p \wedge \Box q)^* = \{\alpha \mid \{p\} \vdash_0 \alpha\} \cap \{\alpha \mid \{q\} \vdash_0 \alpha\}.$$

However,

$$(p \vee q) \notin \Box(p \wedge q) = \{\alpha \mid (p \wedge q)^* \vdash_0 \alpha\} = \{\alpha \mid p^* \cap q^* \vdash_0 \alpha\} = \{\alpha \mid \emptyset \vdash_0 \alpha\}.$$

Thus, under this interpretation, $(\Box p \wedge \Box q) \to \Box(p \wedge q)$ is not true (i.e., $(\Box p \wedge \Box q)^* \not\subseteq (\Box(p \wedge q))^*$). Consult Naumov (2006) for further discussion of this modal logic.

The Logic of Group Decision Making

The final interpretation that I discuss in this section comes from Social Choice Theory. Suppose that $\alpha \in \mathcal{L}_0(\mathsf{At})$ is a propositional formula and I is a set of voters. The interpretation of $\Box \alpha$ explored below is "the group of voters I collectively accept α". In the remainder of this section, I restrict the modal language so that only propositional formulas are within the scope of a modal operator. So, for instance, $\Box \Box p$ is not a well-formed formula. Formally, the modal language of group decision making, denoted $\mathcal{L}^{\flat}(\mathsf{At})$, is the smallest set of formulas generated by the following grammar:

$$\Box \alpha \mid \neg \varphi \mid (\varphi \wedge \psi)$$

where $\alpha \in \mathcal{L}_0(\mathsf{At})$.

The basic idea is that each voter $i \in I$ submits a propositional valuation $v_i : \mathcal{L}_0(\mathsf{At}) \to \{0, 1\}$. The interpretation is that if $v_i(\alpha) = 1$, then the voter i judges α to be true (alternatively, I will say "voter i accepts α"). Given a tuple of propositional valuations (one for each voter),[10] formulas of the form $\Box \alpha$ are interpreted with respect to a group decision-making method. Different ways of aggregating the voters' opinions validate[11] different modal principles. For example, as the reader is invited to check, all the instances (in the language $\mathcal{L}^{\flat}$) of the schemas mentioned in the introduction to this section are valid if $\Box \alpha$ means that there is consensus among the voters in I that α is true. To illustrate, consider axiom (C): If everyone judges that α_1 is true (i.e., for all i, $v_i(\alpha_1) = 1$) and everyone judges that α_2 is true (i.e., for all i, $v_i(\alpha_2) = 1$), then everyone judges that $\alpha_1 \wedge \alpha_2$ is true (i.e., for all i, $v_i(\alpha_1 \wedge \alpha_2) = 1$). Thus, the axiom (C), $\Box \alpha_1 \wedge \Box \alpha_2 \to \Box(\alpha_1 \wedge \alpha_2)$, is valid. However, (C) is not valid if

---

[10] Such a tuple is called a **profile** in the social choice literature.
[11] In this context, a formula $\varphi \in \mathcal{L}^{\flat}$ is **valid** with respect to a group-decision method provided that φ is true given any profile of propositional valuations.

1.3 Why Non-normal Modal Logic?

the the group uses majority rule to make group decisions. The counterexample is the so-called *doctrinal paradox*, which has been extensively discussed in the judgement aggregation literature (List, 2013; Grossi and Pigozzi, 2014). Suppose that there are three voters $I = \{i, j, k\}$ that submit the following valuations:

	p	q	$p \wedge q$
i	1	1	1
j	1	0	0
k	0	1	0
Majority	1	1	0

Then, a majority of the voters accept p and a majority of voters accept q, but only a minority of voters accept $p \wedge q$. That is, $(\Box p \wedge \Box q) \rightarrow \Box(p \wedge q)$ is not true given the above profile of propositional valuations. Thus, $(\Box \alpha_1 \wedge \Box \alpha_2) \rightarrow \Box(\alpha_1 \wedge \alpha_2)$ is not valid when the group uses majority rule to make decisions. Consult Pauly (2007) and Daniëls (2011) for a complete discussion of this interesting interpretation of modal logic.

Exercise 1.10 Find other examples in the literature (or come up with your own!) that motivate interest in non-normal modal logics.

1.4 Why Neighborhood Structures?

In the previous section, I motivated interest in non-normal modal logics by pointing out that there are natural interpretations of the basic modal language that invalidate some theorems and valid rules of any normal modal logic. In this section, I focus on neighborhood structures themselves. The goal is to demonstrate that neighborhood models are an interesting and rich class of mathematical structures that can be fruitfully studied using modal logic. I have already pointed out that many areas of mathematics use sets paired with collections of subsets satisfying certain algebraic properties. Thus, there is general mathematical interest in developing modal languages for reasoning about these specific classes of neighborhood models. The most interesting mathematical structures that fall into this category are topologies. In Sect. 1.4.1, I introduce topological models for modal logic and show how they are related to neighborhood models. Another mathematical structure that is closely related to neighborhood models is a *hypergraph* (Bretto, 2013). In Sect. 1.4.2, I briefly discuss an application of hypergraphs in social choice theory. Nevertheless, general mathematical interest is not the *only* reason to study neighborhood models. Some natural interpretations of neighborhood models make them useful in formal epistemology and game theory. In the remainder of this chapter, I explain how neighborhood models can be used as a semantics for a logic of evidence and belief (Sect. 1.4.4), as a semantics for conditionals (Sect. 1.4.3), or to reason about what players can achieve in game situations (Sect. 1.4.5).

1.4.1 Topological Models

Much of the original motivation for neighborhood structures as a semantics for modal logic comes from elementary point-set topology. In this section, I discuss topological semantics for modal logic. There is a very extensive literature on the topological interpretation of modal logic. It is beyond the scope of this book to discuss all the issues from this literature. Consult van Benthem and Bezhanisvilli (2007); Kremer (2013); Bezhanishvili et al. (2015); Beklemishev and Gabelaia (2014); and Kudinov and Shehtman (2014) for broader surveys and discussions of the main results.

The idea to interpret the basic modal language on topological models is usually attributed to McKinsey and Tarksi (1944). I start by reviewing some concepts from point-set topology. More information can be found in any point-set topology text book (Dugundji 1966, is an excellent choice).

Definition 1.23 (*Topological Space*) A **topological space** is a subset space $\langle W, \mathcal{T} \rangle$, where W is a nonempty set and

1. $W \in \mathcal{T}$ and $\emptyset \in \mathcal{T}$;
2. $\mathcal{T}$ is closed under finite intersections; and
3. $\mathcal{T}$ is closed under arbitrary unions.

Elements $O \in \mathcal{T}$ are called **opens**. A set C such that $W - C \in \mathcal{T}$ is said to be **closed**.

Exercise 1.11 Suppose that $\langle W, \mathcal{T} \rangle$ is a topological space. Prove that the collection of closed sets, $\mathcal{T}_C = \{C \mid W - C \in \mathcal{T}\}$, has the following properties: 1) $W \in \mathcal{T}_C$ and $\emptyset \in \mathcal{T}_C$; 2) $\mathcal{T}_C$ is closed under finite unions; and 3) $\mathcal{T}_C$ is closed under arbitrary intersections.

Suppose that $\langle W, \mathcal{T} \rangle$ is a topological space and $X \subseteq W$ is any set. The largest open subset of X is called the **interior** of X, denoted $Int(X)$. Formally,

$$Int(X) = \bigcup \{O \mid O \in \mathcal{T} \text{ and } O \subseteq X\}.$$

The smallest closed set containing X is called the **closure** of X, denoted $Cl(X)$. Formally,

$$Cl(X) = \bigcap \{C \mid W - C \in \mathcal{T} \text{ and } X \subseteq C\}.$$

It is easy to see that a set X is open if $Int(X) = X$ and closed if $Cl(X) = X$.

Lemma 1.4 *Let $\langle W, \mathcal{T} \rangle$ be a topological space and $X, Y \subseteq W$. Then,*

1. $Int(X \cap Y) = Int(X) \cap Int(Y)$.
2. $Int(\emptyset) = \emptyset$, $Int(W) = W$.
3. $Int(X) \subseteq X$.
4. $Int(Int(X)) = Int(X)$.

1.4 Why Neighborhood Structures?

Exercise 1.12 Suppose that $\langle W, \mathcal{T} \rangle$ is a topological space. Show that for all $X \subseteq W$, $Int(X) = W - Cl(W - X)$. Use this fact to derive properties analogous to those from Lemma 1.4 for $Cl(\cdot)$.

More formally, every topological space $\langle W, \mathcal{T} \rangle$ defines an **interior operator** $Int : \wp(W) \to \wp(W)$ (where for all $X \subseteq W$, $Int(X)$ is defined as above) satisfying the properties from Lemma 1.4 and a closure operator $Cl : \wp(W) \to \wp(W)$ (where for all $X \subseteq W$, $Cl(X)$ is defined as above) satisfying the properties from Exercise 1.12.

Topological spaces can be used as a semantics for a propositional modal language by interpreting the Boolean connectives in the usual way and interpreting the modalities as operators associated with the topology. For instance, McKinsey and Tarski interpret the box-modality as the interior operator for a topological space.

Definition 1.24 (*Topological Model*) A **topological model** is a tuple $\langle W, \mathcal{T}, V \rangle$, where $\langle W, \mathcal{T} \rangle$ is a topology; and $V : \mathsf{At} \to \wp(W)$ is a valuation function.

Suppose that $\mathcal{M}^T = \langle W, \mathcal{T}, V \rangle$ is a topological model. Formulas of $\mathcal{L}(\mathsf{At})$ are interpreted at states $w \in W$. The Boolean connectives and atomic propositions are interpreted as usual. The definition of truth for the modal operator is:

$$\mathcal{M}^T, w \models \Box\varphi \text{ iff there is an } O \in \mathcal{T}, \text{ such that } w \in O \text{ and for all } v \in O,$$
$$\mathcal{M}^T, v \models \varphi.$$

The usual logical notions of validity and satisfiability are defined in the standard way. Recall the notation for the truth set of a formula $\varphi \in \mathcal{L}(\mathsf{At})$: $[\![\varphi]\!]_{\mathcal{M}^T} = \{w \mid \mathcal{M}^T, w \models \varphi\}$. It is an immediate consequence of the definitions that for any formula $\varphi \in \mathcal{L}(\mathsf{At})$ and topological model $\mathcal{M}^T$, $[\![\Box\varphi]\!]_{\mathcal{M}^T} = Int([\![\varphi]\!]_{\mathcal{M}^T})$.

Example 1.25 (*The Usual Topology*) Suppose that $\mathcal{M}^T = \langle \mathbb{R}, \mathcal{T}_{\mathbb{R}}, V \rangle$, where $\mathbb{R}$ is the set of real numbers;

$$\mathcal{T}_{\mathbb{R}} = \{X \mid \text{ for all } x \in X \text{ there is an } \epsilon > 0 \text{ such that } (x - \epsilon, x + \epsilon) \subseteq X\};$$

and $V : \mathsf{At} \to \wp(\mathbb{R})$ is a valuation function with $V(p) = (0, \infty) = \{x \mid x \in \mathbb{R}, x > 0\}$. The topological space $\langle \mathbb{R}, \mathcal{T}_{\mathbb{R}} \rangle$ is called the **usual topology** on $\mathbb{R}$. A set is open in the usual topology if it can be written as the union of open intervals. Since $0 \notin Int([\![p]\!]_{\mathcal{M}^T}) = (0, \infty)$, we have $\mathcal{M}^T, 0 \not\models \Box p$. Furthermore, $[\![\neg p]\!]_{\mathcal{M}^T} = (-\infty, 0] = \{x \mid x \in \mathbb{R} \text{ and } x \leq 0\}$, and so $0 \notin Int([\![\neg p]\!]_{\mathcal{M}^T}) = (-\infty, 0)$. Thus, $\mathcal{M}^T, 0 \not\models \Box\neg p$. Therefore, $\mathcal{M}^T, 0 \not\models \Box p \vee \Box\neg p$.

There are other operators associated with topological spaces that can be used as a semantics for a modal operator. One influential approach is to use the *derived set operator*.[12] Suppose that $\langle W, \mathcal{T} \rangle$ is a topological space. A point $w \in W$ is called a **limit point of** $X \subseteq W$ provided that for each open set $O \in \mathcal{T}$ such that $w \in O$,

---

[12]This interpretation was originally suggested by McKinsey and Tarksi (1944). Consult Bezhanishvili et al. (2010) and Shehtman (1990) for further elaborations of this idea.

$X \cap (O - \{x\}) \neq \emptyset$. The **derived set operator** is a function $Der : \wp(W) \to \wp(W)$, where for all $X \subseteq W$, $Der(X) = \{w \mid w$ is a limit point of $X\}$ ($Der(X)$ is also called the **derivative** of X). The derived set operator is often used as an alternative characterization of closed sets.

Exercise 1.13 Suppose that $\langle W, \mathcal{T} \rangle$ is a topological space. Prove that for any set $X \subseteq W$, $Cl(X) = X \cup Der(X)$.

The key idea is to interpret the diamond modality as the derived set operator. To help keep the two different topological interpretations of the propositional modal language straight, I will use different symbols for the modalities when the diamond operator is interpreted as the derived set operator ($\langle \cdot \rangle$ instead of $\Diamond$ and $\boxdot$ instead of $\Box$). Suppose that $\mathcal{M}^T = \langle W, \mathcal{T}, V \rangle$ is a topological model with $w \in W$. The definition of truth for the two modalities is:

$\mathcal{M}^T, w \models \boxdot \varphi$ iff there is an $O \in \mathcal{T}$ with $w \in O$ and for all $v \in O - \{w\}$, $\mathcal{M}^T, v \models \varphi$.

$\mathcal{M}^T, w \models \langle \cdot \rangle \varphi$ iff for all $O \in \mathcal{T}$ with $w \in O$, there is a $v \in O - \{w\}$ such that $\mathcal{M}^T, v \models \varphi$.

Example 1.26 (*The Usual Topology, again*) Suppose that $\mathcal{M}^T = \langle \mathbb{R}, \mathcal{T}_\mathbb{R}, V \rangle$, where $\mathbb{R}$ is the set of real numbers; $\langle \mathbb{R}, \mathcal{T}_\mathbb{R} \rangle$ is the usual topology (see Example 1.25); and $V : \mathsf{At} \to \wp(\mathbb{R})$ is a valuation function with $V(p) = \{\frac{1}{n} \mid n \geq 1\}$. Then, $\mathcal{M}^T, 0 \models \langle \cdot \rangle p$. That is, $0 \in Der(\llbracket p \rrbracket_{\mathcal{M}^T}) = \llbracket \langle \cdot \rangle \varphi \rrbracket_{\mathcal{M}^T}$. We also have that $\mathcal{M}^T, 0 \not\models \langle \cdot \rangle \langle \cdot \rangle p$. In fact, $Der(\llbracket \langle \cdot \rangle p \rrbracket_{\mathcal{M}^T}) = \emptyset$.

Topological Spaces and Relational Structures

There is a well-known connection between relational frames (Appendix A) and certain topological spaces (see van Benthem and Bezhanisvilli 2007, for a complete discussion). A topological space $\langle W, \mathcal{T} \rangle$ is called an **Alexandroff** space provided that for any (not just finite) $\mathcal{X} \subseteq \mathcal{T}$, $\bigcap \mathcal{X} \in \mathcal{T}$. That is, an Alexandroff topology has the additional property that arbitrary intersections of open sets are open. Now, suppose that $\langle W, R \rangle$ is a reflexive and transitive relational frame—i.e., the relation $R \subseteq W \times W$ is reflexive and transitive. Such a relational frame is called an **S4**-frame. The first observation is that every **S4**-relational frame defines a topological space. A set $X \subseteq W$ is called an R-**upset**, denoted $X^{\uparrow R}$ (or $X^\uparrow$ when it is understood that R is the relation), provided that $w \in X$ and $w R v$ implies that $v \in X$. The set of R-upsets for a reflexive and transitive relation R forms an Alexandroff topology:

Exercise 1.14 Suppose that $\langle W, R \rangle$ is reflexive and transitive relational frame. Let $\langle W, \mathcal{T}_R \rangle$ be a subset space where $\mathcal{T}_R = \{X \mid X$ is an R-upset$\}$. Prove that $\langle W, \mathcal{T}_R \rangle$ is an Alexandroff topology.

We can also construct an **S4**-relational frame from a topology. Suppose that $\langle W, \mathcal{T} \rangle$ is a topological space. The **specialization order**, $R_\mathcal{T} \subseteq W \times W$, is defined as follows $w R_\mathcal{T} v$ iff $v \in Cl(\{w\})$. Thus, $w R_\mathcal{T} v$ provided that v is in every closed set that

1.4 Why Neighborhood Structures?

contains w. It is not hard to see that $\langle W, R_\mathcal{T}\rangle$ is an **S4**-relational frame: It is immediate that $R_\mathcal{T}$ is reflexive. To see that $R_\mathcal{T}$ is transitive, suppose that $v \in Cl(\{w\})$ and $z \in Cl(\{v\})$. Let C be any closed set containing w. Then, since $v \in Cl(\{w\})$, we have $v \in C$. Also, since $v \in C$, C is closed and $z \in Cl(\{v\})$, we have $z \in C$. Thus, z is in every closed set containing w—i.e., $z \in Cl(\{w\})$. Thus, every topological space $\langle W, \mathcal{T}\rangle$ is associated with an **S4**-relational frame $\langle W, R_\mathcal{T}\rangle$. However, while every topology can be associated with an **S4**-relational frame, there is a much tighter connection when the topology is Alexandroff.

Exercise 1.15 Suppose that $\langle W, \mathcal{T}\rangle$ is a topological space. Prove that

- $\mathcal{T} \subseteq \mathcal{T}_{R_\mathcal{T}}$; and
- $\mathcal{T} = \mathcal{T}_{R_\mathcal{T}}$ iff $\mathcal{T}$ is Alexandroff.

Topological Spaces and Neighborhood Structures

Suppose that $\langle W, \mathcal{T}\rangle$ is a topological space. For each $w \in W$, the set of open sets containing w is $\mathcal{T}_w = \{O \mid O \in \mathcal{T} \text{ and } w \in O\}$. A **neighborhood** (in the topological sense) of a point $w \in W$ is a set X such that there is some $O \in \mathcal{T}_w$ such that $O \subseteq X$. That is, X is a neighborhood of w if X contains an open set containing w. For example, in the usual topology on $\mathbb{R}$, the interval $[0, 1]$ is a neighborhood of $\frac{1}{2}$, but it is not a neighborhood of either endpoint (i.e., there is no open set containing 0 contained in $[0, 1]$, and similarly for 1).

Definition 1.27 (Neighborhood System) Suppose that $\langle W, \mathcal{T}\rangle$ is a topology. A **neighborhood system** for $\mathcal{T}$ is a function $N_\mathcal{T}: W \to \wp(\wp(W))$ such that

$$N_\mathcal{T}(w) = \{X \mid \text{there is an } O \in \mathcal{T}_w \text{ such that } O \subseteq X\}$$

Exercise 1.16 Suppose that $\langle W, \mathcal{T}\rangle$ is a topological space. Prove that for all $w \in W$, $N_\mathcal{T}(w)$ is a consistent filter, and that $w \in \bigcap N_\mathcal{T}(w)$.

Neighborhood systems satisfy an additional property that ties together the neighborhoods of different states. Before stating the property, notice that that for all $w \in W$, $\mathcal{T}_w \subseteq N_\mathcal{T}(w)$. That is, any open set containing w is a neighborhood of w (such a set is called an **open neighborhood**).[13] If X is an open neighborhood of w (i.e., $X \in \mathcal{T}_w$), then X is a neighborhood of all of its elements. Thus, any neighborhood system $N_\mathcal{T}$ satisfies the following property:

For all $w \in W$, for all $X \in N_\mathcal{T}(w)$, there is a $Y \subseteq X$ such that for all $v \in Y$, $Y \in N_\mathcal{T}(v)$.

Using Definition 1.27, we have that every $\langle W, \mathcal{T}\rangle$ is associated with a neighborhood frame $\langle W, N_\mathcal{T}\rangle$. It turns out that a class of neighborhood frames well-known to modal logicians gives rise to topological spaces:

---

[13] Typically, $N_\mathcal{T}(w) \not\subseteq \mathcal{T}_w$. That is, in most topologies, there are neighborhoods that are not open.

Definition 1.28 (**S4** *Neighborhood Frame*) A neighborhood frame $\langle W, N \rangle$ is an **S4 neighborhood frame** provided that N satisfies the following properties. For each $w \in W$:

1. $N(w)$ is a consistent filter;
2. $w \in \bigcap N(w)$; and
3. for each $X \subseteq W$, if $X \in N(w)$, then $\{v \mid X \in N(v)\} \in N(w)$.

Proposition 1.1 *Suppose that $\langle W, N \rangle$ is an* **S4**-*neighborhood frame. Then, there is a topology $\langle W, \mathcal{T}_N \rangle$ such that for all $w \in W$, $N(w) = N_{\mathcal{T}_N}(w)$.*

Proof Suppose that $\langle W, N \rangle$ is an **S4**-neighborhood frame. Let $\langle W, \mathcal{T}_N \rangle$ be a subset space where

$$\mathcal{T}_N = \{X \mid \text{for all } w \in W, \text{ if } w \in X, \text{ then } X \in N(w)\}.$$

We first show that $\langle W, \mathcal{T}_N \rangle$ is a topology. Trivially, $\emptyset \in \mathcal{T}_N$. Furthermore, $W \in \mathcal{T}_N$ since, for all $w \in W$, $W \in N(w)$ (this follows from the fact that each $N(w)$ is a consistent filter). Suppose that $O_1, O_2 \in \mathcal{T}_N$ and let $v \in O_1 \cap O_2$. Then, $v \in O_1$ and $v \in O_2$. Hence, $O_1 \in N(v)$ and $O_2 \in N(v)$. Since $N(v)$ is a filter, $O_1 \cap O_2 \in N(v)$; and so, $O_1 \cap O_2 \in \mathcal{T}_N$. Finally, suppose that $\{O_i\}_{i \in I} \subseteq \mathcal{T}_N$ for some index set I. Suppose that $v \in \bigcup_{i \in I} O_i$. Then, $v \in O_i$ for some $i \in I$. Since $O_i \in \mathcal{T}_N$, we have that $O_i \in N(v)$. Therefore, since $O_i \subseteq \bigcup_{i \in I} O_i$ and $N(v)$ is a filter, $\bigcup_{i \in I} \in N(v)$. Thus, $\bigcup_{i \in I} O_i \in \mathcal{T}_N$. This concludes the proof that $\langle W, \mathcal{T}_N \rangle$ is a topology.

To conclude the proof, we show that for all $w \in W$, $N(w) = N_{\mathcal{T}_N}(w)$. Suppose that $w \in W$. If $X \in N_{\mathcal{T}_N}(w)$, then there is some set $O \in \mathcal{T}_N$ such that $w \in O \subseteq X$. Since $O \in \mathcal{T}_N$ and $w \in O$, we have that $O \in N(w)$. Furthermore, since $O \subseteq X$ and $N(w)$ is a filter, we have that $X \in N(w)$. Hence, $N_{\mathcal{T}_N}(w) \subseteq N(w)$. Now, suppose that $X \in N(w)$. We must show that there is some $O \in \mathcal{T}_N$ such that $w \in O \subseteq X$. Let $O = \{v \mid X \in N(v)\}$. Since $X \in N(w)$, $w \in O$. Furthermore, if $v \in O$, then $X \in N(v)$ and $v \in \bigcap N(v) \subseteq X$. Finally, by item 3 of Definition 1.28, since $X \in N(w)$, $O = \{v \mid X \in N(v)\} \in N(w)$. Thus, $X \in N_{\mathcal{T}_N}(w)$, and so, $N(w) \subseteq N_{\mathcal{T}_N}(w)$. Therefore, $N(w) = N_{\mathcal{T}_N}(w)$, as desired. □

Exercise 1.17 Use Proposition 1.1 to prove that for each **S4** neighborhood model $\mathcal{M}$, there is a topological model $\mathcal{M}^T$ such that for all $\varphi \in \mathcal{L}$, $[\![\varphi]\!]_{\mathcal{M}} = [\![\varphi]\!]_{\mathcal{M}^T}$.

Since every topological model can be viewed as an **S4** neighborhood model (i.e., a neighborhood model that satisfies the properties from Definition 1.28), we can say that the class of topological models is *modally equivalent* to the class of **S4**-neighborhood models (cf. Sect. 2.1).

1.4.2 Hypergraphs

A **directed graph** is a pair (W, E) where W is a non-empty set, elements of which are called **nodes** or **vertices**, and $E \subseteq W \times W$, elements of which are called **edges**

1.4 Why Neighborhood Structures?

(cf. the definition of a relational frame in Appendix A). For an *undirected* graph, it is often convenient to let E be a set of subsets of W of size two. In this case, $\{w, v\} \in E$ means that there is an edge between w and v. A **hypergraph** generalizes an undirected graph.[14] Thus, a hypergraph is a subset space $(W, \mathcal{E})$, where $\mathcal{E} \subseteq \wp(W)$ and $\emptyset \notin \mathcal{E}$. The mathematical theory of finite hypergraphs is very well-developed (Bretto, 2013), with applications in combinatorics and optimization problems. Hypergraphs are also used in cooperative game theory and social choice theory, where they are called *simple games* (Taylor and Zwicker, 1999). In this section, I briefly introduce simple games, highlighting an issue that we will return to when discussing the core theory in Chap. 2.

Suppose that I is a finite set of voters. A simple game on I is a monotonic subset space $(I, \mathcal{W})$: $I \neq \emptyset$, $\mathcal{W} \subseteq \wp(I)$ such that for all $U, V \subseteq I$, if $U \in \mathcal{W}$ and $U \subseteq V$, then $V \in \mathcal{W}$. Elements $U \in \mathcal{W}$ are called **winning coalitions**. The intended interpretation is that the set U of voters is a **winning coalition** iff the group selects an option (e.g., the bill or amendment passes, or the candidate is elected) when the voters in U are the ones who voted for it. Given this interpretation, it is clear why it is assumed that all supersets of a winning coalition are winning coalitions. For example, suppose that $I = \{a, b, c, d, e\}$, and consider the following winning coalitions:

$$\mathcal{W} = \{\{d, e\}, \{a, b, c, e\}, \{a, b, d\}, \{b, c, d\}, \{a, c, d\}, \{a, b, c, d\}, \{a, b, c, d, e\}\}.$$

In this example, $\{d, e\} \in \mathcal{W}$ means that, in any voting situation, the group will accept any issue that both d and e agree on. The same is true for the other winning coalitions in $\mathcal{W}$.

An important class of simple games is one that is generated by a *quota rule*. A simple game $(I, \mathcal{W})$ is said to be **weighted** if there is a function $weight : I \to \mathbb{R}$ and quota $q \in \mathbb{R}$, such that for all $U \subseteq I$, $U \in \mathcal{W}$ iff $\sum_{u \in U} weight(u) \geq q$. For instance, the above simple game is generated by the weight function $weight : \{a, b, c, d, e\} \to \mathbb{R}$ where $weight(a) = weight(b) = weight(c) = 1$, $weight(d) = 3$ and $weight(e) = 2$ with the quota $q = 5$. As the reader is invited to verify, the sum of the weights for all the voters in any winning coalition from $\mathcal{W}$ is at least 5.

There is a natural ordering on simple games. Suppose that voters a and b always vote the same way and that voters d and e always vote the same way. In this case, we say that voters $\{a, b\}$ and $\{d, e\}$ form **voting blocs**. It is natural to identify the voters in a voting bloc and treat them as a single voter. Formally, there is an onto function $f : I \to I'$ where $I' = \{a', c, d'\}$ with $f(a) = f(b) = a'$, $f(c) = c$, and $f(d) = f(e) = d'$, that is depicted as follows:

---

[14]There is also a way to define a **directed hypergraph**, although we will not discuss it in this short section.

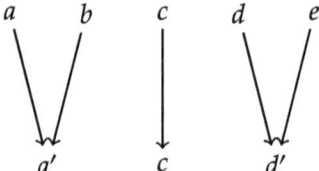

The winning coalitions for the reduced set of voters I' are read off from the original simple game (assuming that the quota is still $q = 5$). For example, $\{d'\}$ is a winning coalition since $f^{-1}[\{d'\}] = \{d, e\} \in \mathcal{W}$. Continuing in this manner, the set of winning coalitions for I' is:

$$\mathcal{W}' = \{\{d'\}, \{a', d'\}, \{c, d'\}, \{a', c, d'\}\}$$

Note that the simple game $(I', \mathcal{W}')$ is a **dictatorship** since there is a single voter $d' \in I'$ such that for all $U \subseteq I'$, $U \in \mathcal{W}'$ iff $d' \in U$. This makes sense since d' represents two voters (d and e) that always vote the same way and whose total weight is 5. This construction is known as the **Rudin–Keisler ordering**, used in the study of ultrafilters.

Definition 1.29 (*Rudin–Keisler Ordering*) Suppose that $\mathcal{G} = (I, \mathcal{W})$ is a simple game. The simple game $\mathcal{G}' = (I', \mathcal{W}')$ is a **RK-projection** of $(I, \mathcal{W})$, denoted $\mathcal{G}' \leq_{RK} \mathcal{G}$, if there is a surjective function $f : I \to I'$ such that for all $X \subseteq I'$, $X \in \mathcal{W}'$ iff $f^{-1}[X] \in \mathcal{W}$. If $\mathcal{G}' \leq_{RK} \mathcal{G}$, then $\mathcal{G}'$ is an RK-**projection of** $\mathcal{G}$.

The Rudin–Keisler ordering is important because it preserves many properties of simple games. For instance, it is not hard to see that $\mathcal{G}' = (I', \mathcal{W}')$ is a weighted simple game (let $weight(a') = 2$, $weight(c) = 1$ and $weight(d') = 5$ with $q = 5$). More generally, it is not hard to see that if $\mathcal{G}$ is a weighted simple game and $\mathcal{G}' \leq_{RK} \mathcal{G}$, then $\mathcal{G}'$ is a weighted simple game. The RK-projection is our first example of a transformation between subset spaces that is intended to preserve important properties. Identifying properties that are preserved by transformations on neighborhood models is an important theme that will be discussed in the next chapter.

Exercise 1.18   1. A simple game $\mathcal{G} = (I, \mathcal{W})$ is called **proper** if for all $X \subseteq I$, if $X \in \mathcal{W}$, then $X^C \notin \mathcal{W}$. Prove that if $\mathcal{G}$ is proper and $\mathcal{G}' = (I', \mathcal{W}') \leq_{RK} \mathcal{G}$, then $\mathcal{G}'$ is also proper.

1.4.3 Conditional Logic

One of the earliest applications of neighborhood models is found in David Lewis's seminal book *Counterfactuals* (1973). In this book, Lewis developed a semantics of conditionals using **sets of spheres**.

1.4 Why Neighborhood Structures?

Definition 1.30 (*Sets of Spheres*) A **set of spheres** is a subset space $\langle W, \mathcal{S} \rangle$, where

- $\mathcal{S}$ is *nested*: For all $S, T \in \mathcal{S}$, either $S \subseteq T$ or $T \subseteq S$.
- $\mathcal{S}$ is *closed under unions*: If $\{S_i \mid i \in I\} \subseteq \mathcal{S}$ for some index set I, then $\bigcup_{i \in I} S_i \in \mathcal{S}$.
- $\mathcal{S}$ is *closed under intersections*: If $\{S_i \mid i \in I\} \subseteq \mathcal{S}$ for some index set I, then $\bigcap_{i \in I} S_i \in \mathcal{S}$.

A system of spheres $\langle W, \mathcal{S} \rangle$ is **centered** on $w \in W$ provided that $\{w\} \in \mathcal{S}$.

Definition 1.31 (*Sphere Frames/Models*) A **sphere frame** is a neighborhood frame $\langle W, N \rangle$, where $W \neq \emptyset$ and for all $w \in W$, $\langle W, N(w) \rangle$ is a set of spheres. We say that $\langle W, N \rangle$ is **centered** provided that for all $w \in W$, $N(w)$ is centered on w.

A **sphere model** is a tuple $\langle W, N, V \rangle$ where $\langle W, N \rangle$ is a sphere frame and $V : \mathsf{At} \to \wp(W)$ is a valuation function.

The idea is that each set in $N(w)$ contains all the states that are "similar" to w to a certain degree. The smaller the sphere (in terms of the subset relation) the more similar the worlds are to w. Lewis (1973, p. 14) explains the intended interpretation of a set of spheres $\langle W, N(w) \rangle$ as follows (I adapt Lewis's notation so that it is consistent with this book):

> Any particular sphere around a world w is to contain just the worlds that resemble w to at least a certain degree. This degree is different for different spheres around w, The smaller the sphere the more similar to w must a world be to fall within it.

The language of conditional logic is a propositional modal language with a binary modality. Formally, the language $\mathcal{L}^{cond}(\mathsf{At})$ is the smallest set of formulas generated by the following grammar:

$$p \mid \neg \varphi \mid (\varphi \wedge \psi) \mid (\varphi \Box\!\!\to \psi)$$

where $p \in \mathsf{At}$ (the set of atomic propositions). The other Boolean connectives ($\vee, \to$, and $\leftrightarrow$) are defined as usual. The dual of the conditional modality, denoted $\varphi \diamondsuit\!\!\to \psi$, is defined as $\neg(\varphi \Box\!\!\to \neg \psi)$. The intended interpretation of $\varphi \Box\!\!\to \psi$ is "if φ, then ψ".

Truth of formulas $\varphi \in \mathcal{L}^{cond}$ is defined at states w from a sphere model $\mathcal{M} = \langle W, N, V \rangle$. I only give the definition of truth for the conditional modality:

$\mathcal{M}, w \models \varphi \Box\!\!\to \psi$ iff either $\bigcup N(w) \cap \llbracket \varphi \rrbracket_\mathcal{M} = \emptyset$ or there is a $S \in N(w)$ such that $\llbracket \varphi \rrbracket_\mathcal{M} \cap S \neq \emptyset$ and $\llbracket \varphi \rrbracket_\mathcal{M} \cap S \subseteq \llbracket \psi \rrbracket_\mathcal{M}$.

Validity and satisfiability are defined in the standard way. The conditional modality $\varphi \Box\!\!\to \psi$ is true at a state w provided that either there is no state in any sphere from $N(w)$ satisfying φ or all the states satisfying φ that are most similar to w also satisfy ψ. Lewis argues that this definition conforms to our intuitions about inference patterns involving conditionals.

Example 1.32 (*Failure of Monotonicity*) The first example highlights a crucial difference between the conditional modality $\varphi \square\!\!\rightarrow \psi$ and the material conditional $\varphi \rightarrow \psi$. It is not hard to see that the material conditional satisfies the following monotonicity property: if $\varphi \rightarrow \psi$ is valid, then so is $(\varphi \wedge \chi) \rightarrow \psi$. However, consider the following example: For many people it is true that "if you put sugar in your coffee, then it will taste good". However, from this statement, one cannot infer that "if you put sugar and gasoline in your coffee, then it will taste good". This is a case in which our intuitions about inferences involving conditionals diverge from valid inference rules involving the material conditional. Lewis's conditional modality does not satisfy this monotonicity property. To illustrate, let $\mathcal{M} = \langle W, N, V\rangle$ be a sphere model with $W = \{w, v_1, v_2, v_3, v_4, v_5, v_6\}$; $N(w) = \{S_1, S_2, S_3, S_4\}$ with $S_1 = \{w\}$, $S_2 = \{w, v_1\}$, $S_3 = \{w, v_1, v_2\}$ and $S_4 = \{w, v_1, v_2, v_3, v_4, v_5, v_6\}$; and $V(p) = \{v_2, v_4, v_5\}$, $V(q) = \{v_1, v_2, v_5, v_6\}$ and $V(r) = \{v_3, v_4\}$. Then, $\mathcal{M}, w \models p\square\!\!\rightarrow q$ since $[\![p]\!]_{\mathcal{M}} \cap S_3 = \{v_2\} \subseteq [\![q]\!]_{\mathcal{M}} = \{v_1, v_2, v_5, v_6\}$; however, $\mathcal{M}, w \not\models (p\wedge r)\square\!\!\rightarrow q$ since S_4 is the only element of $N(w)$ that overlaps $[\![p \wedge r]\!]_{\mathcal{M}}$, but $[\![p \wedge r]\!]_{\mathcal{M}} \cap S_4 = \{v_4\} \not\subseteq [\![q]\!]_{\mathcal{M}}$. The set of spheres $\langle W, N(w)\rangle$ is depicted below (the lined region is $[\![p]\!]_{\mathcal{M}} \cap S_3$ and the grayed region is $[\![p \wedge r]\!]_{\mathcal{M}} \cap S_4$):

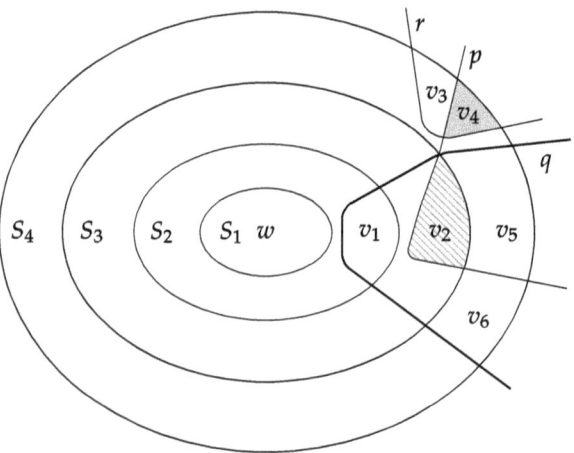

Exercise 1.19 Find a sphere model $\mathcal{M} = \langle W, N, V\rangle$ with a state $w \in W$ such that $p \square\!\!\rightarrow q$ is true at w, $(p \wedge r_1) \square\!\!\rightarrow q$ is false at w, and $(p \wedge (r_1 \wedge r_2)) \square\!\!\rightarrow q$ is true at w.

Example 1.33 (*Failure of Transitivity*) Another property of the material conditional that is not satisfied by the conditional modality is transitivity: $((\varphi \rightarrow \psi) \wedge (\psi \rightarrow \chi)) \rightarrow (\varphi \rightarrow \chi)$ is valid. However, the following example from Stalnaker (1968) illustrates that conditionals should not necessarily satisfy this transitivity property:

1.4 Why Neighborhood Structures?

1. If Hoover had been born a Russian, he would have been a communist.
2. If Hoover were a communist, he would have been a traitor.
3. If Hoover had been born a Russian, then he would have been a traitor.

Intuitively, the first two statements are true; yet, the third is false. To see that the conditional modality does not satisfy this transitivity property, let $\mathcal{M} = \langle W, N, V \rangle$ be a sphere model with $W = \{w, v_1, v_2, v_3, v_4, v_5\}$; $N(w) = \{S_1, S_2, S_3, S_4\}$ with $S_1 = \{w\}$, $S_2 = \{w, v_1\}$, $S_3 = \{w, v_1, v_2\}$ and $S_4 = \{w, v_1, v_2, v_3, v_4, v_5\}$; and $V(p) = \{v_2, v_4\}$, $V(q) = \{v_1, v_2, v_4\}$ and $V(r) = \{w, v_1\}$. Then, $\mathcal{M}, w \models p\,\square\!\!\rightarrow q$ since $[\![p]\!]_\mathcal{M} \cap S_3 = \{v_2\} \subseteq [\![q]\!]_\mathcal{M} = \{v_1, v_2, v_4\}$ and $\mathcal{M}, w \models q\,\square\!\!\rightarrow r$ since $[\![q]\!]_\mathcal{M} \cap S_2 = \{v_1\} \subseteq [\![r]\!]_\mathcal{M} = \{w, v_1\}$; however, $\mathcal{M}, w \not\models p\,\square\!\!\rightarrow r$ since S_3 and S_4 are the only sets that overlap $[\![p]\!]_\mathcal{M}$, but $[\![p]\!]_\mathcal{M} \cap S_3 \not\subseteq [\![q]\!]_\mathcal{M}$ and $[\![p]\!]_\mathcal{M} \cap S_4 \not\subseteq [\![q]\!]_\mathcal{M}$. The set of spheres $\langle W, N(w) \rangle$ is depicted below (the lined region is $[\![p]\!]_\mathcal{M} \cap S_3$ and the grayed region is $[\![q]\!]_\mathcal{M} \cap S_2$):

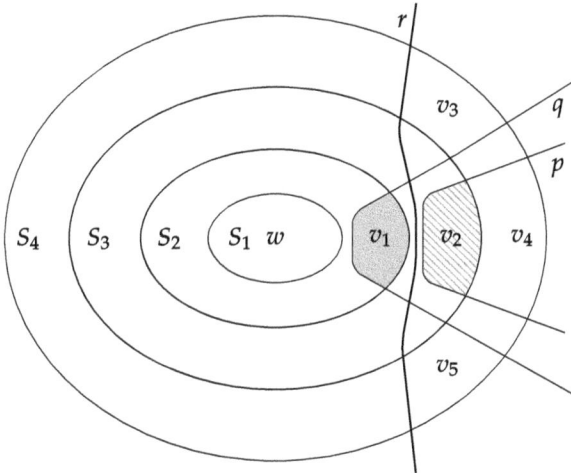

Exercise 1.20 Prove that Modus Tollens is valid for the conditional modality: If $\varphi\,\square\!\!\rightarrow \psi$ and $\neg\psi$ are both valid, then $\neg\varphi$ is valid.

Prove that the rule of contraposition is not valid: Find a sphere model $\mathcal{M} = \langle W, N, V \rangle$ and a state w such that $w \models p\,\square\!\!\rightarrow q$, yet $w \not\models \neg q\,\square\!\!\rightarrow \neg p$.

Exercise 1.21 Find a sphere model $\langle W, N, V \rangle$ with a state $w \in W$ such that $\mathcal{M}, w \not\models p\,\square\!\!\rightarrow q$ and $\mathcal{M}, w \not\models p\,\square\!\!\rightarrow \neg q$.

Remark 1.34 (*Lewis vs. Stalnaker*) Exercise 1.21 shows that *conditional excluded middle*:

$$(\varphi\,\square\!\!\rightarrow \psi) \vee (\varphi\,\square\!\!\rightarrow \neg\psi)$$

is not valid on sphere models. This principle distinguishes Lewis's semantics from a closely related semantics for conditionals proposed by Stalnaker (Stalnaker, 1968; Stalnaker and Thomason, 1970). The crucial observation is that conditional excluded middle is valid on sphere models with an additional constraint:

Stalnaker's condition: Suppose that $\mathcal{M} = \langle W, N, V \rangle$ is a sphere model. If $[\![\varphi]\!]_{\mathcal{M}} \cap \bigcup N(w) \neq \emptyset$, then there is a $S \in N(w)$ such that $S \cap [\![\varphi]\!]_{\mathcal{M}}$ is a singleton.

Thus, according to Stalnaker, $\varphi \square \rightarrow \psi$ is true at a state w provide ψ is true at *the* world v satisfying φ that is most similar to w. A complete comparison between Lewis's and Stalnaker's semantics of conditionals is beyond the scope of this book (see Lewis 1973, Sect. 3.4).

The above examples and exercises highlight important differences between the logic of the conditional modality and the logic of the material conditional. Consult Lewis (1973) for a complete discussion of the main logical and philosophical issues (cf. also Arló-Costa 2007). I conclude this brief introduction to Lewis semantics for conditionals with some remarks about modal languages interpreted on sphere models.

The main observation is that there are other modal operators beyond the conditional modality, $\varphi \square \rightarrow \psi$, that can be used to reason about sphere models. Two natural examples are Lewis's "inner" and "outer" modalities. The outer modality, denoted $[o]\varphi$, describes what is true at any state in any sphere. More formally, the definition of truth at a state w in a sphere model $\mathcal{M} = \langle W, N, V \rangle$ is:

$$\mathcal{M}, w \models [o]\varphi \text{ iff } \bigcup N(w) \subseteq [\![\varphi]\!]_{\mathcal{M}}.$$

Note that the definition of truth for $[o]\varphi$ is the same as the definition of truth of $[\ ]\varphi$ from Sect. 1.2.2. Furthermore, the outer modality is definable using the conditional modality. To see this, note that the following formula is valid:

$$[o]\varphi \leftrightarrow \neg \varphi \square \rightarrow \bot$$

I leave it to the reader to verify the validity of the above formula. Lewis calls the above modality the "outer modality" since it describes what is true in the outermost sphere (since the set of spheres are nested and closed under unions, $\bigcup N(w)$ must be the largest[15] sphere in $N(w)$). The "inner" modality, denoted $[i]\varphi$, describes what is true at all states in the *smallest* sphere (if one exists[16]). The definition of truth for $[i]\varphi$ is essentially the definition of truth for the $\langle\ ]$ modality (see Sect. 1.2.2):

$$\mathcal{M}, w \models [i]\varphi \text{ iff there is some } S \in N(w) \text{ such that } \emptyset \neq S \subseteq [\![\varphi]\!]_{\mathcal{M}}.$$

This modal operator is also definable in terms of the conditional modality. To see this, consider the following formula:

---

[15] That is, "largest" in terms of the subset relation.

[16] Another natural constraint on sphere frames $\langle W, N \rangle$ is the *limit assumption*: for all $w \in W$ and $X \subseteq W$, if $X \cap \bigcup N(w)$, then there is a smallest $Y \in N(w)$ such that $X \cap Y \neq \emptyset$.

1.4 Why Neighborhood Structures?

$$[i]\varphi \leftrightarrow \top \Box\!\!\rightarrow \varphi.$$

As the reader is invited to check, the above formula is valid on all sphere models $\langle W, N, V \rangle$ in which for all $w \in W$, $N(w) \neq \emptyset$.

Lewis's analysis of conditionals uses a second binary modality, denoted $\varphi \preceq \psi$, with the intended interpretation that "φ is *at least as possible* as ψ" or "it is no more far-fetched that φ than that ψ" (Lewis 1973, Sect. 2.5). Truth is defined as follows:

$\mathcal{M}, w \models \varphi \preceq \psi$ iff for all $S \in N(w)$, if $S \cap [\![\psi]\!]_\mathcal{M} \neq \emptyset$, then $S \cap [\![\varphi]\!]_\mathcal{M} \neq \emptyset$.

It is not hard to see that that $\preceq$ is an ordering (i.e., a connected and transitive relation). I will show that $\preceq$ is a connected relation and leave the proof that $\preceq$ is transitive as an exercise. Let $\mathcal{M} = \langle W, N, V \rangle$ be a sphere model with $w \in W$, and suppose that $\mathcal{M}, w \not\models (\varphi \preceq \psi) \vee (\psi \preceq \varphi)$. Then,

1. there is a $S \in N(w)$ such that $S \cap [\![\psi]\!]_\mathcal{M} \neq \emptyset$ and $S \cap [\![\varphi]\!]_\mathcal{M} = \emptyset$; and
2. there is a $S' \in N(w)$ such that $S' \cap [\![\varphi]\!]_\mathcal{M} \neq \emptyset$ and $S' \cap [\![\psi]\!]_\mathcal{M} = \emptyset$.

Since the set of spheres is nested, either $S \subseteq S'$ or $S' \subseteq S$. If $S \subseteq S'$, then by 1, $S' \cap [\![\psi]\!]_\mathcal{M} \neq \emptyset$, which contradicts 2. Similarly, if $S' \subseteq S$, then by 2, $S \cap [\![\varphi]\!]_\mathcal{M} \neq \emptyset$, which contradicts 1. Thus, $\mathcal{M}, w \models (\varphi \preceq \psi) \vee (\psi \preceq \varphi)$; and so $(\varphi \preceq \psi) \vee (\psi \preceq \varphi)$ is valid on all sphere models.

Exercise 1.22 Prove that $(\varphi \preceq \psi) \wedge (\psi \preceq \chi) \rightarrow (\varphi \preceq \chi)$ is valid on all sphere models.

As usual, a "strictly more possible" ordering, denoted $\varphi \prec \psi$, can be defined as $(\varphi \preceq \psi) \wedge \neg(\psi \preceq \varphi)$.[17] Thus, the definition of truth for $\varphi \prec \psi$ is:

$\mathcal{M}, w \models \varphi \prec \psi$ iff there is an $S \in N(w)$ such that $S \cap [\![\varphi]\!]_\mathcal{M} \neq \emptyset$ and $S \cap [\![\psi]\!]_\mathcal{M} = \emptyset$.

An important observation is that relative possibility modality $\prec$ and the conditional modality $\Box\!\!\rightarrow$ are interdefinable. To see this, note that the following formulas are valid on all sphere models:

$$(\varphi \Box\!\!\rightarrow \psi) \leftrightarrow ((\varphi \prec \bot) \rightarrow ((\varphi \wedge \psi) \prec (\varphi \wedge \neg\psi)))$$

$$(\varphi \prec \psi) \leftrightarrow \langle o \rangle (\varphi \vee \psi) \wedge ((\varphi \vee \psi) \Box\!\!\rightarrow \psi)$$

where $\langle o \rangle \chi$ is defined as $\neg [o] \neg \chi$ (which, as shown above, is definable using the conditional modality).

Exercise 1.23 Verify that the above two formulas are valid on all sphere models.

I conclude this section by noting that sphere models can also be used as a semantics for a logic of belief and belief revision (Grove, 1988; Board, 2004; Baltag and Smets,

---

[17] There is also a "equally possible relation", denoted $\varphi \approx \psi$, defined as $(\varphi \preceq \psi) \wedge (\psi \preceq \varphi)$.

2006a). In the literature on modal logics of beliefs and belief revision, the conditional modality $\varphi \,\square\!\!\rightarrow \psi$ is typically denoted $[B]^\varphi \psi$ and represents an agent's *conditional beliefs*. I follow this convention in the subsequent chapters.

1.4.4 A Logic of Evidence and Belief

In this section, I present a logical framework, developed in a series of papers (van Benthem and Pacuit 2011; van Benthem et al. 2012, 2014), in which neighborhood structures are used to represent an agent's beliefs at some fixed moment in time. The key idea is that the neighborhoods at a state represent the set of *evidence* available to the agent at that state. The agent's beliefs are then derived from this set of evidence.

Let W be a set of states (or possible worlds), one of which represents the "actual" world. Assume that there is an agent that gathers *evidence* about this actual state from a variety of sources. To simplify things, assume that these sources provide *binary evidence*—i.e., subsets of W that (may) contain the actual world. There are three assumptions about the agent's source of evidence:

1. Sources may or may not be *reliable*: a piece of evidence (i.e., a subset of worlds) need not contain the actual world. Also, the agent does not know which evidence is reliable.
2. The evidence gathered from different sources (or even the same source) may be jointly inconsistent. Thus, the intersection of all the gathered evidence may be empty.
3. Despite the fact that sources may not be reliable or may be jointly inconsistent, they are all the agent has for forming beliefs.[18]

The *evidential state* of an agent is the set of all propositions identified by the agent's sources. In general, this could be any collection of subsets of W; but there are some minimal constraints:

- no evidence set is empty (evidence per se is never contradictory); and
- the whole universe W is an evidence set (agents know their 'space').

An evidence model is a neighborhood structure satisfying these two constraints:

Definition 1.35 (*Evidence Model*) The tuple $\mathcal{M} = \langle W, E, V \rangle$ is an **evidence model** is a tuple provided that W is a non-empty set of worlds; $E : W \to \wp(\wp(W))$ is a neighborhood function (which is called an **evidence function** in this context); and $V : \text{At} \to \wp(W)$ is a valuation function. Two constraints are imposed on the evidence sets: For each $w \in W$, $\emptyset \notin E(w)$ (E is consistent) and $W \in E(w)$ (E contains the unit).

---

[18] Modeling sources and agents' *trust* in the sources of evidence is also possible, but I will not pursue this in this book.

1.4 Why Neighborhood Structures?

Note that it is not assumed that the collection of evidence sets $E(w)$ is closed under supersets. In addition, an evidence state may contain disjoint sets, whose combination may lead (and should lead) to trouble. But note that, even though an agent may not be able to consistently combine *all* of her evidence, there will be maximal collections of admissible evidence that she can safely put together to form *scenarios*:

Definition 1.36 (*Scenario*) A w-**scenario** is a maximal collection $\mathcal{X} \subseteq E(w)$ that has the fip (i.e., the finite intersection property: for each finite subfamily $\{X_1, \ldots, X_n\} \subseteq \mathcal{X}$, $\bigcap_{1 \leq i \leq n} X_i \neq \emptyset$). A collection is called a **scenario** if it is a w-scenario for some state w.

The modal language used to reason about evidence models uses the modal operator defined in Sect. 1.2.2. Let $\mathcal{L}^{ev}(\mathsf{At})$ be the smallest set of formulas generated by the following grammar:

$$p \mid \neg \varphi \mid (\varphi \wedge \psi) \mid \langle\,]\varphi \mid [B]\varphi \mid [A]\varphi$$

where $p \in \mathsf{At}$. Let $\langle B \rangle$ be defined as $\neg[B]\neg$ and $\langle A \rangle$ be defined as $\neg[A]\neg$. The intended interpretation of $\langle\,]\varphi$ is "the agent has evidence for φ" and $[B]\varphi$ says that "the agent believes that φ is true". The universal modality ($[A]\varphi$: "φ is true in all states") is included for technical convenience. (Alternatively, $[A]\varphi$ can be read as "the agent knows that φ is true".) Truth of formulas in $\mathcal{L}^{ev}(\mathsf{At})$ is defined as follows:

Definition 1.37 (*Truth for Evidence Models*) Let $\mathcal{M} = \langle W, E, V \rangle$ be an evidence model. Truth of a formula $\varphi \in \mathcal{L}^{ev}(\mathsf{At})$ is defined inductively as follows:

- $\mathcal{M}, w \models p$ iff $w \in V(p)$ ($p \in \mathsf{At}$).
- $\mathcal{M}, w \models \neg \varphi$ iff $\mathcal{M}, w \not\models \varphi$.
- $\mathcal{M}, w \models \varphi \wedge \psi$ iff $\mathcal{M}, w \models \varphi$ and $\mathcal{M}, w \models \psi$.
- $\mathcal{M}, w \models \langle\,]\varphi$ iff there exists $X \in E(w)$ such that for all $v \in X$, $\mathcal{M}, v \models \varphi$.
- $\mathcal{M}, w \models [B]\varphi$ iff for each w-scenario $\mathcal{X}$ and for all $v \in \bigcap \mathcal{X}$, $\mathcal{M}, v \models \varphi$.
- $\mathcal{M}, w \models [A]\varphi$ iff for all $v \in W$, $\mathcal{M}, v \models \varphi$.

Recall that the truth set of φ is the set $[\![\varphi]\!]_{\mathcal{M}} = \{w \mid \mathcal{M}, w \models \varphi\}$. The standard logical notions of **satisfiability** and **validity** are defined as usual (cf. Definition 1.13).

According to the above definition, having evidence for φ need not imply that the agent *believes* φ. In order to believe a proposition φ, the agent must consider *all* of her evidence for or against φ. The idea is that each w-scenario represents a maximally consistent theory based on (some of) the evidence collected at w.[19] Note that the definition of truth of the "evidence for" operator builds in monotonicity (recall item 1 in Exercise 1.7). That is, the agent has evidence for φ at w provided that there is some evidence available at w that implies φ.[20]

---

[19] Analogous ideas occur in semantics of conditionals (Kratzer, 1977; Veltman, 1976) and belief revision (Gärdenfors, 1988; Rott, 2001).

[20] Thus, there is a distinction between *having the evidence that* φ (when the truth set of φ is in the agent's set of evidence) and *having evidence for* φ (when there is an evidence set that is contained in the truth set of φ).

The class of evidence models described above is a very general model of an evidential situation. However, there are additional assumptions that can be imposed on an evidence model:

Definition 1.38 (*Flat, Uniform Evidence Models*) An evidence model $\mathcal{M}$ is **flat** if every scenario on $\mathcal{M}$ has non-empty intersection. An evidence model $\mathcal{M} = \langle W, E, V \rangle$ is **uniform** if E is constant. In this case, it is more convenient to treat E as a set (of neighborhoods) rather than as a function.

Proposition 1.2 *The formula $\langle\,]\varphi \to \langle B \rangle \varphi$ is valid on the class of flat evidence models, but not on the class of all evidence models.*

Proof Suppose that $\mathcal{M} = \langle W, E, V \rangle$ is a flat evidence model and that $w \in W$. Suppose that $\mathcal{M}, w \models \langle\,]\varphi$. Then, there is an $X \in E(w)$ such that $X \subseteq [\![\varphi]\!]_\mathcal{M}$. Now, the singleton $\{X\}$ can be extended to a w-scenario $\mathcal{X}_X$.[21] In a flat structure, $\bigcap \mathcal{X}_X \neq \emptyset$; and so, in particular, $[\![\varphi]\!]_\mathcal{M} \cap \bigcap \mathcal{X}_X \neq \emptyset$. Hence, $\mathcal{M}, w \models \langle B \rangle \varphi$. Thus, $\mathcal{M} \models \langle\,]\varphi \to \langle B \rangle \varphi$.

To see that $\langle\,]\varphi \to \langle B \rangle \varphi$ is not valid in general, consider a uniform evidence model $\mathcal{M}_\infty = \langle W, E, V \rangle$ with domain $W = \mathbb{N}$ and evidence sets $E(w) = \mathcal{E} = \{[n, \infty) \mid n \in \mathbb{N}\} \cup \{W\}$ for each $w \in W$. The valuation is unimportant, so we may let $V(p) = \emptyset$ for all $p \in \mathsf{At}$. Clearly, the only scenario on $\mathcal{M}_\infty$ is all of $\mathcal{E}$, but $\bigcap \mathcal{E} = \emptyset$. Hence, $\mathcal{M}_\infty \models [B]\bot$—i.e., $\mathcal{M}_\infty \not\models \langle B \rangle \top$; yet $\mathcal{M}_\infty \models \langle\,]\top$ (this formula is universally valid), and so $\mathcal{M}_\infty \not\models \langle\,]\top \to \langle B \rangle \top$. □

Exercise 1.24 Prove that $(\langle\,]\varphi \wedge [A]\psi) \leftrightarrow \langle\,](\varphi \wedge [A]\psi)$ is valid on all evidence models.

A complete list of axioms for the different classes of evidence models discussed in this section can be found in (van Benthem et al., 2014).

1.4.5 Coalitional Logic

Coalitional logic (Pauly, 2001) uses neighborhood structures to describe the outcomes that (groups of) players can *force* in a game-theoretic situation. Before discussing the logical framework, I need to introduce a few game-theoretic concepts.

Definition 1.39 (*Strategic Game Form*) Suppose that I is a finite non-empty set of players. A **strategic game form** for I is a tuple $\langle I, \{S_i\}_{i \in I}, O, o \rangle$, where for each $i \in I$, S_i is a non-empty set (elements of which are called actions or strategies); O is a non-empty set (elements of which are called **outcomes**); and $o : \Pi_{i \in I} S_i \to O$ is a function assigning an outcome to each tuple of strategies.

---

[21] The formal proof that any singleton can always be extended to a w-scenario uses Zorn's Lemma. I assume that the reader is familiar with Zorn's Lemma and how to use it to prove statements of this form.

1.4 Why Neighborhood Structures?

Elements of $S = \Pi_{i \in I} S_i$ are called **strategy profiles**. Given a strategy profile $s \in S$, let s_i denote i's component and s_{-i} the profile of strategies from s for all players except i. These definitions can be lifted to sets of players, called **coalitions**. So, a strategy profile for a coalition C is a tuple of strategies for each player in C—i.e., an element of $\Pi_{i \in C} S_i$. Let $S_C = \Pi_{i \in C} S_i$ denote the set of strategy profiles for C, and let $\overline{C}$ denote the complement of C (i.e., $\overline{C} = I - C$). Thus, $s_{\overline{C}}$ denotes a profile of strategies, one for each player *not* in C.

Remark 1.40 (Regarding the Definition of a Strategic Game Form)
- A strategic game form can be turned into a game by adding payoffs to each possible outcome (formally, the payoffs are represented by **utility functions** for each player (i.e., for $i \in I$, $u_i : O \to \mathbb{R}$).
- Often, it is assumed that there is exactly one outcome for each profile, so that the set O and $\Pi_{i \in I} S_i$ can be identified. However, for the purposes of presenting the logical framework in this section, it is more convenient to work in a more general setting in which the outcome function o need not be 1-1 or even onto.

Example 1.41 The following game form will be used as a running example in this Section. Let $G_0 = \langle \{A, B\}, \{S_A, S_B\}, O, o \rangle$ be a game in which $S_A = \{s_1, s_2, s_3\}$, $S_B = \{t_1, t_2\}$, $O = \{o_1, o_2, o_3, o_4\}$ and the outcome function o can be read off from the following matrix:

		B	
		t_1	t_2
	s_1	o_1	o_2
A	s_2	o_2	o_3
	s_3	o_4	o_1

The logical system introduced below is intended to describe the outcomes of a game that can be *forced* by a group of players. Consider the example of a strategic game form given above. Player A acting alone cannot force the outcome of the game to be o_1. The best she can do is force the outcome to be one from the set $\{o_1, o_2\}$ by choosing her strategy s_1. Similarly, player B cannot force the game to end in outcome o_1. The best he can do is force to the outcome to be one from the set $\{o_1, o_2, o_4\}$ by choosing strategy t_1. However, as a group, they can force the outcome to be o_1 by choosing their part of the profile $s_{\{A,B\}} = (s_1, t_1)$ (the outcome o_1 can also be arrived at using the strategy profile (s_3, t_2)). This suggests the following definition specifying the sets of outcomes that different coalitions can force.

Definition 1.42 (α-*Effectivity Function*) Suppose that $G = \langle I, \{S_i\}_{i \in I}, O, o \rangle$ is a strategic game form. An α-**effectivity function**[22] is a map $E_G^\alpha : \wp(I) \to \wp(\wp(O))$

---
[22] The term "α-effectivity" comes from the game theory literature (Abdou and Keiding, 1991; Peleg, 1998).

defined as follows: For all $C \subseteq I$, $X \in E_G^\alpha(C)$ iff there exists a strategy profile s_C such that for all $s_{\overline{C}} \in \Pi_{i \in I-C} S_i$, $o(s_C, s_{\overline{C}}) \in X$.

To illustrate, the α-effectivity function $E_{G_0}^\alpha$ for the game form G_0, defined above, is given below. Recall that if $\mathcal{X}$ is a collection of subsets of O, then $\mathcal{X}^{mon}$ is the smallest collection of subsets of O that contains $\mathcal{X}$ and is closed under supersets.

$E_{G_0}^\alpha(\{A\}) = (\{\{o_1, o_2\}, \{o_2, o_3\}, \{o_1, o_4\}\})^{mon}$
$E_{G_0}^\alpha(\{B\}) = (\{\{o_1, o_2, o_4\}, \{o_1, o_2, o_3\}\})^{mon}$
$E_{G_0}^\alpha(\{A, B\}) = (\{\{o_1\}, \{o_2\}, \{o_3\}, \{o_4\}\})^{mon} = \wp(O) - \emptyset$
$E_{G_0}^\alpha(\emptyset) = \{\{o_1, o_2, o_3, o_4\}\}$.

Given any strategic game form G, there is an α-effectivity function E_G^α associated with G. A natural question is: When, exactly, is a function of the form $E : \wp(I) \to \wp(\wp(O))$ for a finite set I of players and a non-empty set of outcomes O the α-effectivity function of some strategic game form G? Naturally, not all functions of the form $E : \wp(I) \to \wp(\wp(O))$ are α-effectivity functions for some game form. Pauly (2001) identified five key properties that characterize α-effectivity functions when there are finitely many outcomes.

1. (*Liveness*) For all $C \subseteq I$, $\emptyset \notin E(C)$.
2. (*Safety*) For all $C \subseteq I$, $O \in E(C)$
3. (*I-maximality*) For all $X \subseteq O$, if $X \in E(I)$ then $\overline{X} \notin E(\emptyset)$.
4. (*Outcome-monotonicity*) For all $X \subseteq X' \subseteq O$, and $C \subseteq I$, if $X \in E(C)$ then $X' \in E(C)$.
5. (*Superadditivity*) For all subsets X_1, X_2 of O and sets of agents C_1, C_2, if $C_1 \cap C_2 = \emptyset$, $X_1 \in E(C_1)$ and $X_2 \in E(C_2)$, then $X_1 \cap X_2 \in E(C_1 \cup C_2)$.

Exercise 1.25 Let $G = \langle I, \{S_i\}_{i \in I}, O, o \rangle$ be a strategic game form and E_G^α the associated α-effectivity functions.

1. Show that E_G^α satisfies all of the above properties. Note that α-effectivity functions satisfy the above properties even if O is infinite.
2. Show that if $E : \wp(I) \to \wp(\wp(O))$ is a function that satisfies superadditivity, then the non-monotonic core of $E(\emptyset)$ is either empty or a singleton.

In a recent paper, Goranko, Jamroga and Turrini (2013) identified the sixth condition that is needed to characterize α-effectivity functions for all outcome sets O. Before stating the condition, I will explain why the above five conditions alone do not characterize α-effectivity functions. The first observation is that in any strategic game form, the empty coalition can force only sets that contain all the possible outcomes in the game. The most elegant way to state this observation uses the notion of the non-monotonic core (cf. Definition 1.1).

Observation 1.43 *Suppose that $G = \langle I, \{S_i\}_{i \in I}, O, o \rangle$ is a strategic game form and E_G^α is the associated α-effectivity function. Then, the non-monotonic core of $E_G^\alpha(\emptyset) = \{range(o)\}$, where $range(o) = \{x \in O \mid \text{there is a } s \in \Pi_{i \in I} S_i \text{ such that } o(s) = x\}$.*

1.4 Why Neighborhood Structures?

The problem is that if E is a function satisfying the above five conditions, then it is possible that the non-monotonic core of the sets that the empty coalition can force may be empty. Of course, this can only happen when O is infinite. The following example from Goranko et al. (2013) (proof of Proposition 4) shows that the above five conditions do not single out all α-effectivity functions: Suppose that there is a single player i with $O = \mathbb{N}$. Consider the function $E : \wp(\{i\}) \to \wp(\wp(O))$ defined as follows:

$E(\{i\}) = \{X \mid X \subseteq \mathbb{N} \text{ is infinite}\};$
$E(\emptyset) = \{X \mid X \subseteq \mathbb{N} \text{ is cofinite (i.e., } \overline{X} \text{ is finite)}\}.$

Since there is no minimal cofinite set, we have $E^{nc}(\emptyset) = \emptyset$. Given Observation 1.43, this means that $E \neq E_G^\alpha$ for any strategic game form G. However, it is not hard to see that E satisfies conditions 1–5. In order to characterize all α-effectivity functions, a sixth condition is needed to rule out examples similar to the one discussed above.

6. (*Empty Coalition*) $E(\emptyset)$ is core complete (cf. Definition 1.3).

I can now state the characterization theorem for α-effectivity functions.

Theorem 1.44 *(Pauly, 2001; Goranko et al., 2013) If $E : \wp(I) \to \wp(\wp(O))$ is a function that satisfies the conditions 1–6 given above, then $E = E_G^\alpha$ for some strategic game form.*

In the remainder of this section, I will introduce a modal logic for reasoning about coalitional powers in games (Pauly, 2001, 2002). Given a finite set of players, or agents, I and a finite or infinite set of atomic propositions At, let $\mathcal{L}^{cl}(\text{At})$ be the smallest set of formulas generated by the following grammar:

$$p \mid \neg \varphi \mid (\varphi \wedge \psi) \mid [C]\varphi$$

where $p \in \text{At}$ and $C \subseteq I$. The other Boolean connectives connectives are defined as usual. The intended interpretation of $[C]\varphi$ is that the players in C have a joint strategy to ensure that φ is true.

Neighborhood structures are used to give a semantics for this language. The set of states W represents the possible outcomes of different strategic game forms. Each state in the model is assigned a function satisfying the above six conditions.

Definition 1.45 (*Coalitional Model*) Suppose that I is a finite set of players. A **coalitional model** for I is a tuple $\langle W, E, V \rangle$, where W is a non-empty set of states $E : W \to (\wp(I) \to \wp(\wp(W)))$ is a function where for all $w \in W$, $E(w)$ satisfies liveness, safety, I-maximality, outcome-monotonicity, superadditivity, and empty coalition; and $V : \text{At} \to \wp(W)$ is a valuation function. To simplify notation, I will write $E(w, C)$ for $E(w)(C)$.

Given a coalitional model $\mathcal{M}$, truth of the formulas from $\mathcal{L}^{cl}(\text{At})$ is defined as in Definition 1.12. I give the definition of truth only for the modal operators:

- $\mathcal{M}, w \models [C]\varphi$ iff $[\![\varphi]\!]_{\mathcal{M}} \in E(w, C)$.

As the reader is invited to check, the following formulas are valid on any coalitional model:

$$
\begin{aligned}
\text{(Liveness)} \quad & \neg[C]\bot \\
\text{(Safety)} \quad & [C]\top \\
\text{(I − maximality)} \quad & [I]\varphi \rightarrow \neg[\emptyset]\neg\varphi \\
\text{(Monotonicity)} \quad & [C](\varphi \wedge \psi) \rightarrow [C]\varphi \wedge [C]\psi \\
\text{(Superadditivity)} \quad & ([C_1]\varphi_1 \wedge [C_2]\varphi_2) \rightarrow [C_1 \cup C_2](\varphi_1 \wedge \varphi_2) \\
& \text{provided that } C_1 \cap C_2 = \emptyset.
\end{aligned}
$$

Each of the formulas corresponds[23] to the five properties needed to characterize α-effectivity functions. The sixth property (*empty coalition*) does not have a corresponding modal axiom. Nonetheless, (Pauly, 2001, 2002) proved that these axioms are sound and complete for the class of coalitional models. See (Goranko et al., 2013) for a complete discussion.

---

[23] I am using "*correspondence*" in the sense of modal correspondence theory, which is discussed in Sect. 2.5.

Chapter 2
Core Theory

The previous chapter established that neighborhood structures with the basic propositional modal language is an interesting and well-motivated logical framework. In this chapter, I move away from questions of motivation to explore the logical theory of neighborhood structures.

The main object of study is a neighborhood model $\langle W, N, V \rangle$ in which W is a non-empty set; N assigns a collection of subsets of W to each state; and V assigns a subset of W to each atomic proposition (see Definitions 1.11 and 1.9). In order to facilitate a comparison with relational models, it is convenient to let N be a relation $N \subseteq W \times \wp(W)$ (cf. Remark 1.10). Two different definitions of truth for the modal operator can be found in the literature. In order to compare and contrast these two definitions, I introduced two different modalities (here, I give the definition of truth treating N as a relation):

- $\mathcal{M}, w \models \langle\,]\varphi$ iff there is a $X \subseteq W$ such that $w\ N\ X$ and $X \subseteq [\![\varphi]\!]_\mathcal{M}$.
- $\mathcal{M}, w \models \Box\varphi$ iff there is a $X \subseteq W$ such that $w\ N\ X$ and $X = [\![\varphi]\!]_\mathcal{M}$.

These two modalities are equivalent when the neighborhoods are monotonic (i.e., so that if $w\ N\ X$ and $X \subseteq Y$, then $w\ N\ Y$; see the discussion in Sect. 1.2.2). It is clear from this presentation that neighborhood models generalize the standard relational models $\langle W, R, V \rangle$, where $R \subseteq W \times W$ for the basic modal language (cf. Appendix A). Indeed, much of the mathematical theory of modal logic with respect to relational structures can be adapted to the more general setting involving neighborhood structures. In particular, there is a well-behaved notion of structural equivalence between neighborhood models matching the expressivity of the basic modal language (Sect. 2.1); there is a well-developed proof theory for weak systems of modal logic (Sects. 2.3 and 2.4.3); the canonical model method for proving axiomatic completeness can be adapted to the more general setting (Sect. 2.3.2); there is a generalization of frame correspondence theory linking properties of the neighborhood relation and valid formulas (Sect. 2.5); the satisfiability problem for non-normal modal logics is decidable (Sect. 2.4.1); and there is a standard translation into first-order logic (Sect. 2.6.3).

However, there are some important differences between neighborhood semantics and relational semantics for modal logic. Two of the most striking properties are that the satisfiability problem for many non-normal modal logics is **NP**-complete as opposed to **PSPACE**-complete (Sect. 2.4.2), and that there are consistent normal modal logics that are incomplete with respect to relational semantics but complete with respect to neighborhood semantics (Sect. 2.3.3). Finally, an important theme in this chapter is the relationship between neighborhood models and other semantics for the basic modal language (Sect. 2.2).

2.1 Expressive Power and Invariance

Once a language and semantics are defined, the first steps towards a model theory is to identify an appropriate notion of *structural equivalence* between models matching the expressivity of the language. For example, the appropriate notion of structural equivalence for first-order logic is an *isomorphism* (Enderton 2001, Chap. 2). For the basic modal language $\mathcal{L}$, the appropriate notion of structural equivalence between relational models (Definition A.1) is a *bisimulation* (Definition A.9). In this section, I show that there is a natural notion of a bisimulation between neighborhood models.

I start with the definition of *modal equivalence*. Suppose that $\mathcal{M}$ is a neighborhood model. I write $dom(\mathcal{M})$ for the **domain** of $\mathcal{M}$—i.e., the set of states in $\mathcal{M}$. A pair $\mathcal{M}, w$ with $w \in dom(\mathcal{M})$ is called a **pointed model**. For each pointed model $\mathcal{M}, w$, let

$$Th_{\mathcal{L}}(\mathcal{M}, w) = \{\varphi \in \mathcal{L} \mid \mathcal{M}, w \models \varphi\}.$$

The set of formulas $Th_{\mathcal{L}}(\mathcal{M}, w)$ is called the **theory** of $\mathcal{M}, w$—i.e., the set of all modal formulas true at w in $\mathcal{M}$. If $Th_{\mathcal{L}}(\mathcal{M}, w) = Th_{\mathcal{L}}(\mathcal{M}', w')$, then the two situations $\mathcal{M}, w$ and $\mathcal{M}', w'$ are indistinguishable from the point of view of the modal language $\mathcal{L}$.

Definition 2.1 ($\mathcal{L}$-*Equivalence*) Suppose that $\mathcal{M}, w$ and $\mathcal{M}', w'$ are two pointed neighborhood models and $\mathcal{L}$ is a modal language. We say that $\mathcal{M}, w$ and $\mathcal{M}', w'$ are $\mathcal{L}$-**equivalent**, denoted $\mathcal{M}, w \equiv_{\mathcal{L}} \mathcal{M}', w'$, when $Th_{\mathcal{L}}(\mathcal{M}, w) = Th_{\mathcal{L}}(\mathcal{M}', w')$. If $\mathcal{L}$ is the basic modal language (Definition 1.6) and $\mathcal{M}, w \equiv_{\mathcal{L}} \mathcal{M}', w'$, then we say that $\mathcal{M}, w$ and $\mathcal{M}', w'$ are modally equivalent.

Exercise 2.1 Use Exercise 1.7 part 2 to show that for all monotonic pointed models $\mathcal{M}, w$ and $\mathcal{M}', w'$ and the language $\mathcal{L}^{mon}$ from Sect. 1.2.2, $\mathcal{M}, w \equiv_{\mathcal{L}} \mathcal{M}', w'$ iff $\mathcal{M}, w \equiv_{\mathcal{L}^{mon}} \mathcal{M}', w'$.

There is a natural notion of bisimulation between *monotonic* neighborhood models. In order to facilitate a comparison with the definition of a bisimulation on relational models, I state the following definition treating the neighborhood functions as relations.

2.1 Expressive Power and Invariance

Definition 2.2 (*Monotonic Bisimulation*) Suppose that $\mathcal{M} = \langle W, N, V \rangle$ and $\mathcal{M}' = \langle W', N', V' \rangle$ are two monotonic neighborhood models. A relation $Z \subseteq W \times W'$ is a **monotonic bisimulation** provided that, whenever wZw':

Atomic harmony: for each $p \in \mathsf{At}$, $w \in V(p)$ iff $w' \in V'(p)$.
Zig: If $w\ N\ X$, then there is an $X' \subseteq W'$ such that $w'\ N'\ X'$ and $\forall x' \in X', \exists x \in X$ such that $x\ Z\ x'$.
Zag: If $w'\ N'\ X'$, then there is an $X \subseteq W$ such that $w\ N\ X$ and $\forall x \in X, \exists x' \in X'$ such that $x\ Z\ x'$.

Write $\mathcal{M}, w \underline{\leftrightarrow} \mathcal{M}', w'$ when there is a monotonic bisimulation $Z \subseteq dom(\mathcal{M}) \times dom(\mathcal{M}')$ such that $w\ Z\ w'$.

A simple, but instructive, induction on the structure formulas shows that monotonic bisimulations preserve truth over models:

Proposition 2.1 *Suppose that Z is a monotonic bisimulation between two monotonic models $\mathcal{M} = \langle W, N, V \rangle$ and $\mathcal{M}' = \langle W', N', V' \rangle$. Then, for all $\varphi \in \mathcal{L}$, for all $w \in W$, $w' \in W'$, if wZw', then $\mathcal{M}, w \models \varphi$ iff $\mathcal{M}', w' \models \varphi$. That is, $\mathcal{M}, w \underline{\leftrightarrow} \mathcal{M}', w'$ implies that $\mathcal{M}, w \equiv_\mathcal{L} \mathcal{M}', w'$.*

It is well known that (relational) bisimulations (Definition A.9) completely characterize modal equivalence on certain classes of relational models. For instance, for all *image-finite relational models* (relational models such that for all $w \in W$ the set of states accessible from w is finite), two states are modally equivalent iff the states are bisimilar. An analogous result holds on monotonic neighborhood models. Before stating this result, it is convenient to restrict the definition of a monotonic bisimulation to *non-monotonic core* of the neighborhood models (Definition 1.1). More formally, a **monotonic core bisimulation** is similar to a monotonic bisimulation, except that the zig and zag clauses are restricted to the non-monotonic core. For instance, the zig-condition of a monotonic core bisimulation is:

Zig^{nc}: If $X_1 \in N_1^{nc}(w_1)$, then there is an $X_2 \subseteq W_2$ such that $X_2 \in N_2^{nc}(w_2)$ and $\forall x_2 \in X_2, \exists x_1 \in X_1$ such that $x_1\ Z\ x_2$.

A key observation is that on core-complete monotonic models, every monotonic core bisimulation is a monotonic bisimulation, and vice versa.

Proposition 2.2 *Suppose that $\mathcal{M}_1$ and $\mathcal{M}_2$ are core-complete monotonic neighborhood models. Then, Z is a monotonic bisimulation between $\mathcal{M}_1$ and $\mathcal{M}_2$ iff Z is a monotonic core bisimulation.*

Exercise 2.2 Prove Proposition 2.2.

I can now define a class of models for which there is a perfect match between bisimilarity and modal equivalence.

Definition 2.3 (*Locally Core-Finite*) A neighborhood model $\mathcal{M} = \langle W, N, V \rangle$ is locally core-finite provided that $\mathcal{M}$ is core-complete and for each $w \in W$, $N^{nc}(w)$ is finite, and for all $X \in N^{nc}(w)$, X is finite.

Obviously, any model with finitely many states is locally core-finite. However, a model with infinitely many states can still be locally core-finite. I first prove that there is a perfect match between bisimilarity and modal equivalence on finite monotonic neighborhood models.

Theorem 2.4 *Suppose that $\mathcal{M} = \langle W, N, V \rangle$ and $\mathcal{M}' = \langle W', N', V' \rangle$ are finite monotonic models (i.e., W and W' are finite sets). Then, for all $w \in W$, $w' \in W'$, $\mathcal{M}, w \equiv_{\mathcal{L}} \mathcal{M}', w'$ iff $\mathcal{M}, w \underline{\leftrightarrow} \mathcal{M}', w'$.*

Proof The right-to-left direction is Proposition 2.1 (the result holds for all monotonic neighborhood models). For the left-to-right direction, suppose that $\mathcal{M} = \langle W, N, V \rangle$ and $\mathcal{M}' = \langle W', N', V' \rangle$ are monotonic locally core-finite models. We show that modal equivalence $\equiv_{\mathcal{L}}$ is a monotonic bisimulation (to simplify the notation, write $\equiv$ instead of $\equiv_{\mathcal{L}}$).

Suppose that $X \in N(w)$. We must show that there exists $X' \in N'(w')$ such that for all $x' \in X'$, there exists $x \in X$ such that $x \equiv x'$. Suppose not. Since both models are finite, we have $N'(w') = \{X'_1, \ldots, X'_k\}$ and $X = \{x_1, \ldots, x_m\}$. Thus, the assumption is that for each $i = 1, \ldots, k$, there exists $x'_i \in X'_i$ such that $(*)$ for all $x_j \in X$, $x_j \not\equiv x'_i$. Fix a set of elements $x'_i \in X'_i$ for $i = 1, \ldots, k$ satisfying $(*)$. This means that for each $i = 1, \ldots, k$, for each $j = 1, \ldots, m$, there is a formula φ_{ij} such that $\mathcal{M}, x_j \models \varphi_{ij}$, but $\mathcal{M}', x'_i \not\models \varphi_{ij}$. Now, we have $\mathcal{M}, x_j \models \bigwedge_{i=1,\ldots,k} \varphi_{ij}$; and so,

$$X \subseteq [\![\bigvee_{j=1,\ldots,m} \bigwedge_{i=1,\ldots,k} \varphi_{ij}]\!]_{\mathcal{M}}$$

Let $\varphi := \bigvee_{j=1,\ldots,m} \bigwedge_{i=1,\ldots,k} \varphi_{ij}$. Then, $\mathcal{M}, w \models \Box\varphi$. Since $\mathcal{M}, w \equiv \mathcal{M}', w'$, we have $\mathcal{M}', w' \models \Box\varphi$. However, this is a contradiction, since there is no $i = 1, \ldots, k$ such that $X'_i \subseteq [\![\varphi]\!]_{\mathcal{M}'}$. $\square$

Exercise 2.3 Use Theorem 2.4 and Proposition 2.2 to prove that monotonic bisimulations characterize modal expressivity on locally core-finite monotonic neighborhood models:

Theorem 2.5 *Suppose that $\mathcal{M} = \langle W, N, V \rangle$ and $\mathcal{M}' = \langle W', N', V' \rangle$ are monotonic, locally core-finite models. Then, for all $w \in W$, $w' \in W'$, $\mathcal{M}, w \equiv_{\mathcal{L}} \mathcal{M}', w'$ iff $\mathcal{M}, w \underline{\leftrightarrow} \mathcal{M}', w'$.*

Theorem 2.5 shows that monotonic bisimulations capture modal expressivity on locally core-finite monotonic neighborhood models.[1] Interestingly, the above notion of a bisimulation applies only when the models are monotonic.

---

[1]This result can be generalized to the class of *modally saturated* neighborhood models (Hansen 2003; Hansen et al. 2009).

Example 2.6 (*Monotonic Bisimulations*) Suppose that $\mathcal{M}$ and $\mathcal{M}'$ are neighborhood models, the first of which is not monotonic. Consider the relation Z pictured below between the domains of $\mathcal{M}$ and $\mathcal{M}'$.

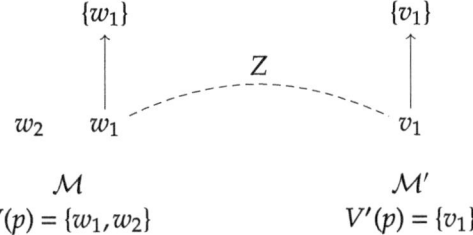

As the reader is invited to check, the dashed line satisfies all the conditions in Definition 2.2. That is, Z is a monotonic bisimulation. However, $\mathcal{M}, w_1 \not\models \Box p$ and $\mathcal{M}', v_1 \models \Box p$.

Thus, if the neighborhood models are not closed under supersets, then monotonic bisimulations do not necessarily preserve the truth of modal formulas.

Exercise 2.4 Prove that monotonic bisimulations preserves the truth of the modal language $\mathcal{L}^{mon}$ on neighborhood models even if they are not monotonic. This suggests that it is smoother to develop a model theory for neighborhood models using the language $\mathcal{L}^{mon}$ (cf. Hansen 2003).

The above example raises a question: What is the right notion of equivalence between *arbitrary* neighborhood models? A complete answer to this question is discussed in Hansen et al. (2009). The main idea is to use *bounded morphisms*.[2]

Definition 2.7 (*Bounded Morphism*) Suppose that $\mathcal{M}_1 = \langle W_1, N_1, V_1 \rangle$ and $\mathcal{M}_2 = \langle W_2, N_2, V_2 \rangle$ are two neighborhood models. If $f : W_1 \to W_2$ and $X \subseteq W_2$, then let $f^{-1}(X) = \{w \in W_1 \mid f(w) \in X\}$ be the inverse image of X. A function $f : W_1 \to W_2$ is a **bounded morphism** if

1. for all $p \in \mathsf{At}$, $w \in V_1(p)$ iff $f(w) \in V_2(p)$; and
2. for all $X \subseteq W'$, $f^{-1}(X) \in N_1(w)$ iff $X \in N_2(f(w))$.

Bounded morphisms preserve the truth of the basic modal language.

Proposition 2.3 *Suppose that* $\mathcal{M} = \langle W, N, V \rangle$ *and* $\mathcal{M}' = \langle W', N', V' \rangle$ *are two neighborhood models, and* $f : W \to W'$ *is a bounded morphism. Then, for all* $\varphi \in \mathcal{L}$, $\mathcal{M}, w \models \varphi$ *iff* $\mathcal{M}', f(w) \models \varphi$.

Proof The proof is by induction on the structure of φ. The argument for the base case and Boolean connectives is as usual. We only give the argument for the modal operator. Suppose that φ is of the form $\Box \psi$.

---

[2] The analogue of a bounded morphism for relational models is a *p-morphism* (Blackburn et al. 2001, Sect. 2.1).

The induction hypothesis is that for all $w \in W$, $\mathcal{M}, w \models \varphi$ iff $\mathcal{M}', f(w) \models \varphi$. This means that $f^{-1}(\llbracket \varphi \rrbracket_{\mathcal{M}'}) = \llbracket \varphi \rrbracket_{\mathcal{M}_1}$ (where $f^{-1}(X) = \{w \mid f(w) \in X\}$ is the inverse image of X). Then,

$\mathcal{M}, w \models \Box \psi$ iff $\llbracket \psi \rrbracket_{\mathcal{M}} \in N(w)$ (definition of truth)
    iff $f^{-1}(\llbracket \psi \rrbracket_{\mathcal{M}'}) = \llbracket \psi \rrbracket_{\mathcal{M}} \in N(w)$ (induction hypothesis)
    iff $\llbracket \psi \rrbracket_{\mathcal{M}'} \in N'(f(w))$ (definition of bounded morphism)
    iff $\mathcal{M}', f(w) \models \Box \psi$ (definition of truth)

$\Box$

Exercise 2.5 What is the relationship between the Rudin–Keisler ordering discussed in Sect. 1.4.2 and bounded morphisms? For further comparisons between the Rudin–Keisler ordering and monotonic bisimulations, consult, Daniëls (2011).

Remark 2.8 (*Bounded Morphisms vs. Monotonic Bisimulations*) Note that the relation between the models in Example 2.6 is actually a function from W' to W (strictly speaking, the converse of the relation Z is the function). Let $f : W' \to W$ be the function where $f(v_1) = w_1$. It is instructive to see why this function is not a *bounded morphism* from $\mathcal{M}'$ to $\mathcal{M}$. The second condition of Definition 2.7 requires that for all $X \subseteq W'$, $f^{-1}(X) \in N(w)$ iff $X \in N'(f(w))$. Let $X = \{w_1, w_2\}$. Then, $X \notin N'(f(v_1))$. However, $f^{-1}(X) = \{w_1\} \in N(w_1)$. Thus, f is not a bounded morphism. Note, also, that Z is not a bounded morphism from $\mathcal{M}$ to $\mathcal{M}'$ since a bounded morphism must be a total function.

I start with an illustrative example. Consider the following two neighborhood models: $\mathcal{M} = \langle W, N, V \rangle$ and $\mathcal{M}' = \langle W', N', V' \rangle$. In the first model, $W = \{w_1, w_2, w_3\}$, $N(w_1) = N(w_2) = \{w_2\}$ and $N(w_3) = \emptyset$. In the second model, $W' = \{v\}$ and $N(v) = \emptyset$. In both models, all propositional variables are false at all states. The models are pictured below. Note that w_1, w_2 and v are modally equivalent. At all three states, all formulas of the form $\Box \varphi$ are false and all formulas of the form $\Diamond \varphi$ are true. Since $N'(v) = \emptyset$, but $N(w_1) = N(w_2) \ne \emptyset$, there is no monotonic bisimulation between $\mathcal{M}$ and $\mathcal{M}'$. Rather than trying to find a relationship between the two models, the idea is to show that both models can "live" inside a third model in such a way that modally equivalent states can be identified. For example, let $\mathcal{N} = \langle W'', N'', V'' \rangle$ be a neighborhood model with $W'' = \{s_1, s_2\}$, $N''(s_1) = \{\emptyset\}$, $N''(s_2) = \emptyset$, and for all atomic propositions p, $V''(p) = \emptyset$. There are bounded morphisms $f : \mathcal{M} \to \mathcal{N}$ and $g : \mathcal{M}' \to \mathcal{N}$ such that $f(w_1) = f(w_2) = g(v)$. The models and bounded morphisms are pictured below:

2.1 Expressive Power and Invariance

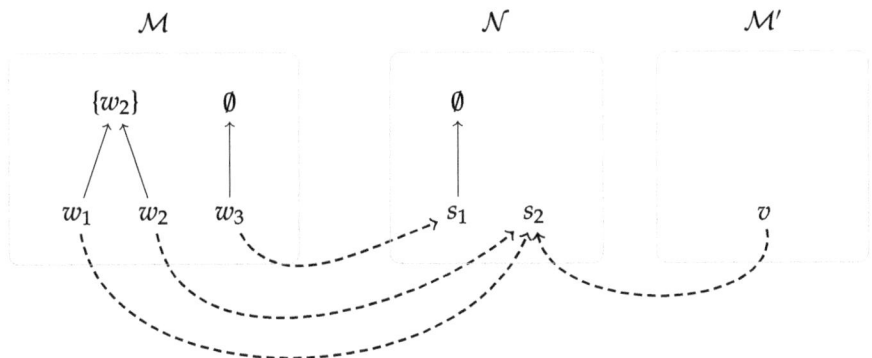

Definition 2.9 (*Behavioral Equivalence*) Suppose that $\mathcal{M} = \langle W, N, V \rangle$ and $\mathcal{M}' = \langle W', N', V' \rangle$ are neighborhood models, and let $w \in W$ and $w' \in W'$. Then, $\mathcal{M}, w$ and $\mathcal{M}', w'$ are **behaviorally equivalent** iff there is a neighborhood model $\mathcal{N} = \langle W'', N'', V'' \rangle$ and bounded morphisms $f : \mathcal{M} \to \mathcal{N}$ and $g : \mathcal{M}' \to \mathcal{N}$ such that $f(w) = g(w')$.

Proposition 2.4 *Suppose that $\mathcal{M} = \langle W, N, V \rangle$ and $\mathcal{M}' = \langle W', N', V' \rangle$ are two neighborhood models. If states $w \in W$ and $w' \in W'$ are behaviorally equivalent, then for all $\varphi \in \mathcal{L}$, $\mathcal{M}, w \models \varphi$ iff $\mathcal{M}', w' \models \varphi$.*

The proof is an immediate consequence of Proposition 2.3.

Disjoint Unions

An important feature of the modal language is that the definition of truth is "local". This feature is best exemplified by the fact that taking the *disjoint union* of models does not affect the truth of formulas at states in each component.

Definition 2.10 (*Disjoint Union*) Let $\{\langle W_i, N_i, V_i \rangle\}_{i \in I}$ be a collection of neighborhood models with disjoint sets of states. The **disjoint union** is the model $\uplus_i \mathcal{M}_i = \langle W, N, V \rangle$, where $W = \bigcup_{i \in I} W_i$; for all $p \in \mathsf{At}$, $V(p) = \bigcup_{i \in I} V_i(p)$; and

$$\text{For all } X \subseteq \bigcup_{i \in I} W_i, \ X \in N(w) \text{ iff } X \cap W_i \in N_i(w).$$

Proposition 2.5 *Suppose that for all $i \in I$, $\mathcal{M}_i = \langle W_i, N_i, V_i \rangle$. Then, for each $i \in I$ and $w \in W_i$, for all $\varphi \in \mathcal{L}$, $\mathcal{M}_i, w \models \varphi$ iff $\uplus_i \mathcal{M}_i, w \models \varphi$.*

The above proposition can be directly proved by an induction on the structure of formulas in $\mathcal{L}$. Alternatively, one can show that the natural embedding of each model into the disjoint union is a bounded morphism.

Exercise 2.6 Suppose that $\{\mathcal{M}_i = \langle W_i, N_i, V_i \rangle\}_{i \in I}$ is a collection of neighborhood models with disjoint sets of states and that $\uplus_i \mathcal{M}_i = \langle W, N, V \rangle$ is the disjoint union of the models. Prove that for each $i \in I$ and $w \in W_i$, for all the natural embedding of each model $\mathcal{M}_i$ into the disjoint union $\uplus_i \mathcal{M}_i$ (the 1-1 function that embeds each W_i into $\bigcup_i W_i$) is a bounded morphism.

Comparing Different Classes of Models

An important theme in this book is to compare and contrast neighborhood models with alternative models for the basic modal language. Often, the goal is to show that certain classes of neighborhood models are *modally equivalent* to some other class of models. In order to express this more formally, I need some notation. Suppose that M is a class of models (such as relational models), and that pM is the resulting class of pointed models—i.e., pairs $\mathcal{M}, w$ where $\mathcal{M}$ is a model from M and w is a state from $\mathcal{M}$. Each class of models M comes with a definition of *truth* for the basic modal language $\mathcal{L}$. Formally, "truth" for the modal language $\mathcal{L}$ is a relation, denoted $\models_M$, between pointed models from pM and formulas $\varphi \in \mathcal{L}$ (write $\mathcal{M}, w \models \varphi$ when φ is true at w in $\mathcal{M}$). The definition of modal equivalence between neighborhood models (Definition 2.1) can be generalized to this more general setting.

Definition 2.11 (*Modal Equivalence between Classes of Models*) Suppose that $\mathcal{L}$ is a modal language, and M and M' are two classes of models for $\mathcal{L}$. Let $\mathcal{M}, w$ be a pointed models from pM and $\mathcal{M}', w'$ be a pointed model from pM'. Say that $\mathcal{M}, w$ is $\mathcal{L}$-equivalent to $\mathcal{M}', w'$, denoted $\mathcal{M}, w \equiv_{\mathcal{L}} \mathcal{M}', w'$, provided that $Th_{\mathcal{L}}(\mathcal{M}, w) = \{\varphi \mid \mathcal{M}, w \models_M \varphi\} = \{\varphi \mid \mathcal{M}', w' \models_{M'} \varphi\} = Th_{\mathcal{L}}(\mathcal{M}', w')$. If $\mathcal{L}$ is the basic modal language, then we say that $\mathcal{M}, w$ and $\mathcal{M}, w'$ are **modally equivalent**.

A class of models M is $\mathcal{L}$-equivalent to a class of models M' provided for each pointed model $\mathcal{M}, w$ from pM, there exists a pointed model $\mathcal{M}', w'$ from pM' such that $\mathcal{M}, w \equiv_{\mathcal{L}} \mathcal{M}', w'$, and *vice versa*.

Typically, demonstrating that M and M' are modally equivalent involves showing how to transform models from M into models from M' and, conversely, how to transform models from M' into models from M. For instance, the following theorem is a direct consequence of Proposition 1.17 from Sect. 1.4.1.

Theorem 2.12 *The class* $T = \{\mathcal{M}^T \mid \mathcal{M}^T$ *is a topological model* $\}$ *is modally equivalent to the class* $M_{S4} = \{\mathcal{M} \mid \mathcal{M}$ *is an* **S4** *neighborhood model*$\}$.

2.2 Alternative Semantics for Non-normal Modal Logics

Neighborhood models are not the only semantics for the basic modal language. Indeed, depending on the intended interpretation of the modalities, neighborhood models may not always be the best choice of semantics for weak modal logics (cf. the discussion of logics of ability in Sect. 1.3). It is important to understand the relationship between neighborhood models and alternative semantics for the basic modal language. To keep the discussion manageable, in this section, I focus on variations of relational models (Definition A.1). Consult Venema (2007) and Chellas

(1980, Exercises 7.11, 7.42, 7.43, and 8.33) for discussions of *algebraic models*. There are also *coalgebraic* models for the basic modal language (Kupke and Pattinson 2011) that generalize neighborhood models (Hansen and Kupke 2004; Hansen et al. 2009; Venema 2007).

2.2.1 Relational Models

Let R be a relation on a non-empty set W (i.e., $R \subseteq W \times W$). For each $w \in W$, let $R(w) = \{v \mid wRv\}$, and for each $X \subseteq W$, let $R(X) = \{w \mid \exists v \in X \text{ such that } wRv\}$. So, $R(w)$ is the set of states that w can "see" via the relation R, and $R(X)$ is the set of states that can "see" some element of X (via the relation R).

Definition 2.13 (*R-Necessity*) Let R be a relation on a non-empty set W and $w \in W$. A set $X \subseteq W$ is **R-necessary at** w if $R(w) \subseteq X$. Let $\mathcal{N}_w^R$ be the set of sets that are R-necessary at w (we simply write $\mathcal{N}_w$ if R is clear from the context). That is, $\mathcal{N}_w^R = \{X \mid R(w) \subseteq X\}$.

The following Lemma shows that the collection of R-necessary sets for some relation R have very nice algebraic properties.

Lemma 2.1 *Let R be a relation on W. Then, for each $w \in W$, $\mathcal{N}_w$ is augmented.*

Exercise 2.7 Prove Lemma 2.1.

Furthermore, properties of R are reflected in this collection of sets.

Observation 2.14 *Let W be a set and $R \subseteq W \times W$.*

1. *If R is reflexive, then for each $w \in W$, $w \in \bigcap \mathcal{N}_w$*
2. *If R is transitive, then for each $w \in W$, if $X \in \mathcal{N}_w$, then $\{v \mid X \in \mathcal{N}_v\} \in \mathcal{N}_w$.*

Proof Suppose that R is reflexive. Let $w \in W$ be an arbitrary state. Suppose that $X \in \mathcal{N}_w$. Then, since R is reflexive, wRw and, hence, $w \in R(w)$. Therefore by the definition of $\mathcal{N}_w$, $w \in X$. Since X was an arbitrary element of $\mathcal{N}_w$, $w \in X$ for each $X \in \mathcal{N}_w$. Hence, $w \in \bigcap \mathcal{N}_w$.

Suppose that R is transitive. Let $w \in W$ be an arbitrary state. Suppose that $X \in \mathcal{N}_w$. We must show $\{v \mid X \in \mathcal{N}_v\} \in \mathcal{N}_w$. That is, we must show $R(w) \subseteq \{v \mid X \in \mathcal{N}_v\}$. Let $x \in R(w)$. Then wRx. To complete the proof, we need only show that $X \in \mathcal{N}_x$. That is, we must show $R(x) \subseteq X$. Since R is transitive, $R(x) \subseteq R(w)$ (why?). Hence, since $R(w) \subseteq X$, $R(x) \subseteq X$. □

Exercise 2.8 State and prove analogous results for the situations in which R is serial (for all $w \in W$, there exists a v such that wRv), Euclidean (for all $w, v, u \in W$, if wRv and wRu then vRu) and symmetric (for all $w, v \in W$, if wRv, then vRw).

Recall that a **relational frame** $\mathfrak{F}$ is a tuple $\langle W, R \rangle$, where $W \neq \emptyset$ and $R \subseteq W \times W$; and a **relational model** for $\mathcal{L}(\mathsf{At})$ based on $\mathfrak{F}$ is a tuple $\langle \mathfrak{F}, V \rangle$ where $V : \mathsf{At} \to \wp(W)$ is a propositional valuation function (cf. Definition A.1). Both relational models and neighborhood models can be used to provide a semantics for the basic modal language (cf. Appendix A). It should be clear that neighborhood models are more general than relational models (that is, neighborhood models satisfy more sets of formulas than relational models). The following Theorem identifies the class of neighborhood models that is modally equivalent to the class of relational models.

Theorem 2.15 *The class* $\mathsf{K} = \{\mathfrak{M} \mid \mathfrak{M}$ *is a relational model* $\}$ *is modally equivalent to the class* $\mathsf{M}_{aug} = \{\mathcal{M} \mid \mathcal{M}$ *is an augmented neighborhood model*$\}$.

The proof of this Theorem starts with a definition of equivalence between neighborhood and relational frames.

Definition 2.16 Let W be a nonempty set of states, $\langle W, N \rangle$ a neighborhood frame and $\langle W, R \rangle$ a relational frame. We say that $\langle W, N \rangle$ and $\langle W, R \rangle$ are **point-wise equivalent** provided that for all $X \subseteq W$, $X \in N(w)$ iff $X \in \mathcal{N}_w^R$.

Exercise 2.9 Prove that if a neighborhood frame $\mathcal{F} = \langle W, N \rangle$ and a relational frame $\mathfrak{F} = \langle W, R \rangle$ are point-wise equivalent, then for any propositional valuation $V : \mathsf{At} \to \wp(W)$, if $\mathcal{M} = \langle \mathcal{F}, V \rangle$ and $\mathfrak{M} = \langle \mathfrak{F}, V \rangle$, then for all $w \in W$, $\mathcal{M}, w$ and $\mathfrak{M}, w$ are modally equivalent.

Using Exercise 2.9, the proof of Theorem 2.15 is a simple consequence of the following two Lemmas.

Lemma 2.2 *Let* $\langle W, R \rangle$ *be a relational frame. Then, there is a modally equivalent augmented neighborhood frame.*

Proof The proof is straightforward given Lemma 2.1: for each $w \in W$, let $N(w) = \mathcal{N}_w^R$ (where R is the relation under consideration). □

Lemma 2.3 *Let* $\langle W, N \rangle$ *be an augmented neighborhood frame. Then, there is a modally equivalent relational frame.*

Proof Let $\langle W, N \rangle$ be a neighborhood frame. We must define a relation R_N on W. Since $\langle W, N \rangle$ is augmented, for each $w \in W$, $\cap N(w) \in N(w)$. For each $w, v \in W$, let $w R_N v$ iff $v \in \cap N(w)$. To show that $\langle W, R_N \rangle$ and $\langle W, N \rangle$ are equivalent, we must show that for each $w \in W$, $\mathcal{N}_w^{R_N} = N(w)$. Let $w \in W$ and $X \subseteq W$. If $X \in \mathcal{N}_w^{R_N}$, then $R_N(w) \subseteq X$. Since $R_N(w) = \cap N(w)$ and N contains its core, $R_N(w) \in N(w)$. Furthermore, since N is supplemented and $R_N(w) = \cap N(w) \subseteq X$, $X \in N(w)$. Now, suppose that $X \in N(w)$. Then, clearly, $\cap N(w) \subseteq X$. Hence, $X \in \mathcal{N}_w^{R_N}$. □

Exercise 2.10 Suppose that K_{eq} is the class of relational models $\mathfrak{M} = \langle W, R, V \rangle$, where R is an equivalence relation. Find the class of neighborhood models that is modally equivalent to K_{eq}.

2.2.2 Generalized Relational Models

The next class of models is intended to provide a natural semantics for so-called *non-adjunctive logics*. These are modal logics that do not include the axiom scheme (C): $(\Box\varphi \land \Box\psi) \to \Box(\varphi \land \psi)$. Schotch and Jennings introduced a semantics for such logics in a series of papers (Schotch and Jennings 1980; Jennings and Schotch 1981, 1980).

Definition 2.17 (*n-ary Relational Model*) An *n*-ary relational model (where $n \geq 2$) is a tuple $\langle W, R, V \rangle$, where W is a non-empty set and $R \subseteq W^n$ is an *n*-ary relation,[3] and $V : \mathsf{At} \to \wp(W)$ is a valuation function.

So, relational models (cf. Definition A.1) are 2-ary models. The definition of truth for the basic modal language $\mathcal{L}(\mathsf{At})$ follows the usual pattern. Let $\mathfrak{M}^n = \langle W, R, V \rangle$ be an *n*-ary relational model and $w \in W$. The Boolean connectives are defined as usual. The clauses for the modal operators are:

- $\mathfrak{M}^n, w \models \Box\varphi$ iff for all $(w_1, \ldots, w_{n-1}) \in W^{n-1}$, if $(w, w_1, \ldots, w_{n-1}) \in R$, then there exists i such that $1 \leq i \leq n$ and $\mathfrak{M}^n, w_i \models \varphi$.
- $\mathfrak{M}^n, w \models \Diamond\varphi$ iff there exists $(w_1, \ldots, w_{n-1}) \in W^{n-1}$ such that $(w, w_1, \ldots, w_{n-1}) \in R$, and for all i such that $1 \leq i \leq n$, we have $\mathfrak{M}^n, w_i \models \varphi$.

An *n*-ary frame is a pair $\langle W, R \rangle$, where $R \subseteq W^n$ is an *n*-ary relation. The standard logical notions of satisfiability and validity are defined as usual (cf. Definition 1.13).

Example 2.18 (*A 3-ary Relational Model*) Let $\mathfrak{M}^3 = \langle W, R, V \rangle$ be a 3-ary relational model for the modal language generated from the atomic propositions $\mathsf{At} = \{p, q, r\}$, where $W = \{w_1, w_2, w_3, w_4, w_5, w_6, w_7\}$; $R = \{(w_1, w_2, w_3), (w_1, w_4, w_5), (w_1, w_6, w_7)\}$; and $V(p) = \{w_2, w_4, w_6\}$, $V(q) = \{w_3, w_5, w_7\}$ and $V(r) = \{w_2, w_3, w_7\}$. This model is depicted as follows:

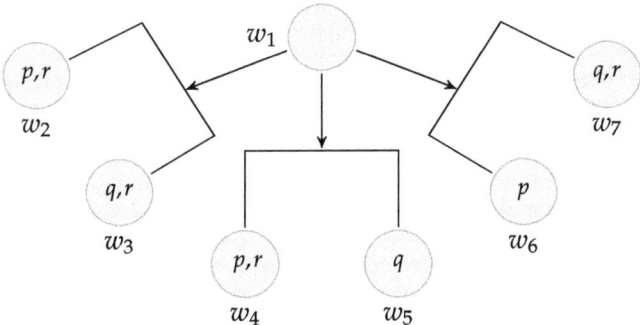

According to the above definition of truth for the modal operators on *n*-ary relational models, we have:

---

[3]I write X^n for the *n*-fold cross product of the set X. That is, X^n consists of all tuples $\langle x_1, \ldots, x_n \rangle$ of length n where each $x_i \in X$.

- $\mathfrak{M}^3, w_1 \models \Box p$ (and $\mathfrak{M}^3, w_1 \models \Box \neg p$);
- $\mathfrak{M}^3, w_1 \models \Box q$ (and $\mathfrak{M}^3, w_1 \models \Box \neg q$); and
- $\mathfrak{M}^3, w_1 \not\models \Box(p \wedge q)$.

The above model shows that the axiom scheme (C) is not valid on the class of 3-ary relational models. Consider, again, the model $\mathfrak{M}^3$ given in Example 2.18. Note that $\mathfrak{M}^3, w_1 \models \Box r$, so we have $\mathfrak{M}^3 \models \Box p \wedge \Box q \wedge \Box r$. While $\Box$ is not "closed under conjunction",[4] a weaker conjunctive closure condition is satisfied: $\mathfrak{M}^3, w_1 \models \Box((p \wedge r) \vee (q \wedge r))$. The primary interest in n-ary relational models is that they can be used to study a hierarchy of weaker conjunctive closure principles. For each $n \geq 2$, define the following formula:

$$(C^n) \quad \bigwedge_{i=1}^{n} \Box \varphi_i \to \Box \bigvee_{1 \leq k,l \leq n,\ k \neq l} (\varphi_k \wedge \varphi_l).$$

So, for example, C^3 is the formula

$$(\Box \varphi_1 \wedge \Box \varphi_2 \wedge \Box \varphi_3) \to \Box((\varphi_1 \wedge \varphi_2) \vee (\varphi_2 \wedge \varphi_3) \vee (\varphi_1 \wedge \varphi_3)).$$

Exercise 2.11 Prove that the formula C^3 is valid on 3-ary relational frames.

Allen (2005) showed that every finite n-ary relational structure is modally equivalent to a finite monotonic neighborhood structure, and vice versa (cf. Arló-Costa 2005).

Theorem 2.19 *The class* $\mathsf{K}^n = \{\mathfrak{M}^n \mid \mathfrak{M}^n \text{ is an } n\text{-ary relational model}\}$ *is modally equivalent to the class*

$$\mathsf{M}_{mon} = \{\mathcal{M} \mid \mathcal{M} \text{ is a non-trivial monotonic neighborhood model}\}.$$

To illustrate, I give an example showing how to translate a neighborhood model into a modally equivalent n-ary relational model.

Example 2.20 (Neighborhoods to n-ary Relations) Let $\mathcal{M} = \langle W, N, V \rangle$ be a monotonic neighborhood model with $W = \{w, v\}$; $N(w) = \{\{w\}, \{v\}, \{w, v\}\}$ and $N(v) = \{\{w, v\}\}$; and $V(p) = \{w\}$ and $V(q) = \{v\}$. Note that $N^{nc}(w) = \{\{w\}, \{v\}\}$ and $N^{nc}(v) = \{\{w, v\}\}$. The first step is to add copies of the states so that each minimal neighborhood contains exactly two sets. To that end, let $\mathcal{M}' = \langle W', N', V' \rangle$ with $W' = W \cup \{w', v'\}$, where w' is a copy of w and v' is a copy of v. So, $N'(v) = N'(v') = \{\{w, v\}, \{w', v'\}\}$ and $N'(w) = N'(w') = \{\{w\}, \{v\}\}$; and $V'(p) = \{w, w'\}$ and $V'(q) = \{v, v'\}$. The second step is to construct a 3-ary relational model $\mathcal{M}^{N'} = \langle W^{N'}, R^{N'}, V^{N'} \rangle$ in which

- $W^{N'} = \{w, v, w', v'\}$;
- $R^{N'} = \{(w, w, v), (v, w, w'), (v, w, v'), (v, v, w'), (v, v, v')\}$; and
- $V^{N'} = V'$.

---
[4]In fact, we have $\mathfrak{M}^3, w_1 \not\models \Box(p \wedge q)$, $\mathfrak{M}^3, w_1 \not\models \Box(p \wedge r)$, and $\mathfrak{M}^3, w_1 \not\models \Box(q \wedge r)$.

Exercise 2.12 Prove the following proposition:

Proposition 2.6 *Suppose that $\mathcal{M} = \langle W, N, V \rangle$ is a finite monotonic neighborhood model such that for all $w \in W$, $N(w) \neq \emptyset$. Then, there is an n-ary relational model $\mathfrak{M}^N = \langle W^N, R^N, V^N \rangle$ that is modally equivalent to $\mathcal{M}$.*

The proof that every finite n-ary relational model can be transformed into a monotonic neighborhood model is more involved (consult Allen 2005).

2.2.3 Multi-relational Models

Goble (2000) used models with a set of relations as a semantics for a deontic logic in which there are possibly conflicting obligations arising from different normative sources (cf., also, Governatori and Rotolo 2005).

Definition 2.21 (*Multi-Relational Model*) Suppose that At is a set of atomic propositions. A **multi-relational model** is a triple $\langle W, \mathcal{R}, V \rangle$, where W is a non-empty set; $\mathcal{R} \subseteq \wp(W \times W)$ is a set of serial relations (i.e., for all $R \in \mathcal{R}$, for all $w \in W$, there exists $v \in W$ such that $w \mathrel{R} v$); and $V : \text{At} \to \wp(W)$ is a valuation function.

The definition of truth for the basic modal language $\mathcal{L}(\text{At})$ follows the usual pattern. Let $\mathfrak{M} = \langle W, \mathcal{R}, V \rangle$ be a multi-relational model and $w \in W$. Boolean connectives are defined as usual. The clauses for the modal operators are:

- $\mathfrak{M}, w \models \Box \varphi$ iff there exists $R \in \mathcal{R}$ such that for all $v \in W$, if $w \mathrel{R} v$, then $\mathfrak{M}, v \models \varphi$.
- $\mathfrak{M}, w \models \Diamond \varphi$ iff for all $R \in \mathcal{R}$ there is a $v \in W$ such that $w \mathrel{R} v$ and $\mathfrak{M}, v \models \varphi$.

Note that, according to the above definition, the relations in a multi-relational model $\langle W, \mathcal{R}, V \rangle$ are assumed to be serial. This means that for all states $w \in W$, for all $R \in \mathcal{R}$, the set $R(w) = \{v \mid w \mathrel{R} v\}$ is non-empty. This assumption can be dropped, but doing so will lead to some complications. A state $w \in W$ is said to be a dead-end state with respect to a relation R provided that $R(w) = \emptyset$ (i.e., there are no states accessible from w). This means that if w is a dead-end state for a relation $R \in \mathcal{R}$ in a multi-relational model $\mathfrak{M}$, then $\mathfrak{M}, w \models \Box \bot$. When studying non-adjunctive logics, it is important to distinguish between situations in which $\Box \varphi \wedge \Box \neg \varphi$ is true and situations in which $\Box \bot$ is true. Ruling out dead-end states ensures that $\neg \Box \bot$ is valid.

2.2.4 Impossible Worlds

Impossible worlds were first introduced into modal logic by Saul Kripke (1965) to provide a semantics for some historically important systems of modal logic weaker than **K**.[5] Impossible worlds can be used in a variety of ways to weaken systems of modal logic (see Berto 2013 for a discussion). I will briefly discuss how to use impossible worlds to provide a semantics for regular modal logics. These are modal logics in which the necessitation rule is not valid (equivalently, modal logics that do not contain $\Box\top$).

Say that a world w is an **impossible world** in a model $\mathfrak{M}$ if nothing is necessary (no formulas of the form $\Box\varphi$ are true at $\mathfrak{M}, w$) and everything is possible (all formulas of the form $\Diamond\varphi$ are true at $\mathfrak{M}, w$). The key idea is to distinguish between possible and impossible worlds in a relational model.

Definition 2.22 (*Impossible worlds*) A **relational model with impossible worlds** is a tuple $\langle W, W_N, R, V \rangle$, where W is a non-empty set of worlds; $W_N \subseteq W$; $R \subseteq W \times W$; and $V : \mathsf{At} \to \wp(W)$.

Suppose that $\mathfrak{M} = \langle W, W_N, R, V \rangle$ is a relational model with impossible worlds. Truth for the basic modal language is defined as usual, except for the modal clause:

- $\mathfrak{M}, w \models \Box\varphi$ iff $w \in W_N$ and for all $v \in W$, if $w\ R\ v$, then $\mathfrak{M}, v \models \varphi$.

Adding impossible worlds to relational models is an elegant way to invalidate the necessitation rule (while keeping all other axioms and rules of normal modal logic intact). Consider any atomic proposition p, and suppose that $\mathfrak{M} = \langle W, W_N, R, V \rangle$ is a relational model with impossible worlds consisting of worlds $W = \{w, v\}$ and $W_N = \{w\}$ with $R = \{(w, v)\}$. Since the interpretation of the Boolean connectives is as usual at both possible and impossible worlds, we have that $p \vee \neg p$ is valid on any relational model with impossible worlds (in particular, $\mathfrak{M} \models p \vee \neg p$). However, since $v \notin W_N$, we have $\mathfrak{M}, v \not\models \Box(p \vee \neg p)$. Thus, we have $\mathfrak{M}, w \models \Box(p \vee \neg p)$; yet, since $w\ R\ v$, we have $\mathfrak{M}, w \not\models \Box\Box(p \vee \neg p)$. Thus, (Nec) is not valid over the class of relational models with impossible worlds.

There is much more to say about impossible worlds and how they can be used to model various non-normal modal logics. The interested reader is invited to consult Priest (2008) and Berto (2013), and references therein, for a more extensive discussion.

Exercise 2.13 Suppose that $\mathfrak{M} = \langle W, W_N, R, V \rangle$ is a relational model with impossible worlds. Find a neighborhood model $\mathcal{M}$ that is modally equivalent to $\mathfrak{M}$.

2.3 The Landscape of Non-normal Modal Logics

I argued in Sect. 1.3 that there is interest in studying so-called *non-normal modal logics*. These are weak systems of modal logics in which one or more of the following formulas and rules are not valid.

---

[5]Note that Kripke called impossible worlds "non-normal".

2.3 The Landscape of Non-normal Modal Logics

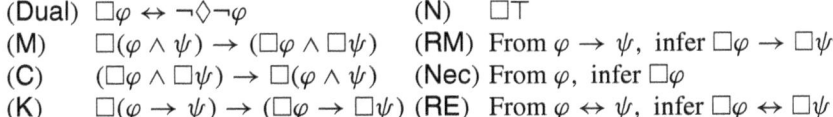

(Dual) $\Box\varphi \leftrightarrow \neg\Diamond\neg\varphi$ (N) $\Box\top$
(M) $\Box(\varphi \wedge \psi) \to (\Box\varphi \wedge \Box\psi)$ (RM) From $\varphi \to \psi$, infer $\Box\varphi \to \Box\psi$
(C) $(\Box\varphi \wedge \Box\psi) \to \Box(\varphi \wedge \psi)$ (Nec) From φ, infer $\Box\varphi$
(K) $\Box(\varphi \to \psi) \to (\Box\varphi \to \Box\psi)$ (RE) From $\varphi \leftrightarrow \psi$, infer $\Box\varphi \leftrightarrow \Box\psi$

There are two natural questions to ask about the above formulas and rules. First, which of them are valid on all neighborhood models? Second, are all the formulas and rules *independent*? That is, which of the axioms or rules can be *derived* from the others?

I start with the first question. There are neighborhood models that invalidate each of (M), (C), (K), (N), (RM) and (Nec). Example 1.14 is a countermodel to an instance of (M) (and also (RM)— see Lemma 2.6 below).

Observation 2.23 *There are formulas φ and ψ such that* (C), $(\Box\varphi \wedge \Box\psi) \to \Box(\varphi \wedge \psi)$, *and* (K), $\Box(\varphi \to \psi) \to (\Box\varphi \to \Box\psi)$, *are not valid on the class of all neighborhood frames.*

Proof For the first formula, consider the neighborhood model $\mathcal{M} = \langle W, N, V \rangle$ with $W = \{w, v\}$, $N(w) = \{\{w\}, \{v\}\}$, $N(v) = \{\emptyset\}$ and $V(p) = \{w\}$ and $V(q) = \{v\}$. Thus, $\mathcal{M}, w \models \Box p \wedge \Box q$, but since $[\![p \wedge q]\!]_\mathcal{M} = \emptyset \notin N(w)$, $\mathcal{M}, w \not\models \Box(p \wedge q)$.

For the second formula, we construct the following neighborhood model: $\mathcal{M} = \langle W, N, V \rangle$ with $W = \{w, v, s\}$, $N(w) = \{\{w\}, \{w, v, s\}\}$, $V(p) = \{w\}$ and $V(q) = \{w, v\}$. Then, $[\![p]\!]_\mathcal{M} = \{w\}$, $[\![q]\!]_\mathcal{M} = \{w, v\}$ and $[\![p \to q]\!]_\mathcal{M} = (\neg p \vee q)^\mathcal{M} = \{w, v, s\}$. Thus, we have $\mathcal{M}, w \models \Box(p \to q) \wedge \Box p$ but $\mathcal{M}, w \not\models \Box q$. $\square$

Exercise 2.14 Can you find a neighborhood model with a state in which all $\Box$-formulas are false, but all $\Diamond$-formulas are true? Is this possible with relational semantics?

So, which formulas *are* valid on all neighborhood frames? As noted above, Definition 1.12 ensures that $\Box$ and $\Diamond$ are duals. This gives the following validity:

Lemma 2.4 *The schema* (Dual), $\Box\varphi \leftrightarrow \neg\Diamond\neg\varphi$, *is valid on any neighborhood frame.*

Proof Let $\mathcal{M} = \langle W, N, V \rangle$ be any neighborhood model. Then, using the Definition of truth, the above properties of truth sets, and basic set-theoretic reasoning, we have the following equivalences for any $w \in W$:

$$\begin{aligned}
\mathcal{M}, w \models \Box\varphi \ &\text{iff}\ [\![\varphi]\!]_\mathcal{M} \in N(w) \\
&\text{iff}\ W - (W - [\![\varphi]\!]_\mathcal{M}) \in N(w) \\
&\text{iff}\ W - ([\![\neg\varphi]\!]_\mathcal{M}) \in N(w) \\
&\text{iff}\ \mathcal{M}, w \not\models \Diamond\neg\varphi \\
&\text{iff}\ \mathcal{M}, w \models \neg\Diamond\neg\varphi
\end{aligned}$$

Thus, $\Box\varphi \leftrightarrow \neg\Diamond\neg\varphi$ is valid. □

In addition, the inference rule (RE) is valid on the class of neighborhood frames.

Lemma 2.5 *On the class of all neighborhood frames, if $\varphi \leftrightarrow \psi$ is valid, then $\Box\varphi \leftrightarrow \Box\psi$ is valid.*

Proof The simple (and instructive!) proof is left to the reader. □

The proof that (Dual) and (RE) are the only axiom and rules (in addition to propositional logic) needed to axiomatize the class of all neighborhood structures can be found in Sect. 2.3.2.

Let us now turn to the second question: Which axioms/rules can be *derived* from the others? I assume that the reader is familiar with the basics of Hilbert-style axiomatizations of modal logics. See Appendix A.2 for a brief introduction and Blackburn et al. (2001, Sect. 1.6) for a more extensive discussion. Let (PC) denote any axiomatization of propositional logic and (MP) denote the rule of Modus Ponens (from φ and $\varphi \to \psi$ infer ψ). The smallest (in the sense of the least number of consequences) logical system that we study in this book is **E**:

- **E** is the smallest set of formulas containing all instances of (PC) and (Dual) and is closed under the rules (RE) and (MP).

The other logical systems will be extensions of **E**. For example, the logic **EC** is the smallest set of formulas containing all instances of (PC), (Dual) and (C), and is closed under the rules (RE) and (MP). That is, **EC** extends the logic **E** by adding all instances of the axiom scheme (C). This is also the case for **EM**, **EN**, **ECM**, **EK** and **EMCN**. The logic **K** is the smallest set of formulas containing all instances of (PC), (Dual), (K), and the rules (Nec) and (MP). Note the difference between **K** and **EK**. Let **L** be any of the above logics; $\vdash_\mathbf{L} \varphi$ means that $\varphi \in \mathbf{S}$, and, in such a case, φ is said to be a **theorem of L**. As usual, if $\vdash_\mathbf{L} \varphi$, then there is a **deduction** of φ in the logic **L**.

Definition 2.24 (*Tautology*) A modal formula φ is called a **tautology** if $\varphi = (\alpha)^\sigma$ where σ is a substitution, α is a formula of propositional logic and α is a tautology.

For example, $\Box p \to (\Diamond(p \wedge q) \to \Box p)$ is a tautology because $a \to (b \to a)$ is a tautology in the language of propositional logic and

$$(a \to (b \to a))^\sigma = \Box p \to (\Diamond(p \wedge q) \to \Box p)$$

where $\sigma(a) = \Box p$ and $\sigma(b) = \Diamond(p \wedge q)$.

Definition 2.25 (*Deduction*) Suppose that **L** is an extension of **E**. A **deduction** in **L** is a finite sequence of formulas $\alpha_1, \ldots, \alpha_n$ where for each $i = 1, \ldots, n$ either (1) α_i is a tautology; (2) α_i is an instance of the axioms of **L**; or (3) α_i follows by Modus Ponens or the other rules of **L** from earlier formulas.[6]

---

[6] For Modus Ponens this means that there is $j, k < i$ such that α_k is of the form $\alpha_j \to \alpha_i$.

2.3 The Landscape of Non-normal Modal Logics

It is useful to introduce some terminology to classify different systems of propositional modal logic.

Definition 2.26 (*Classifying Modal Logics*) A propositional modal logic **L** is called

- a **normal modal logic** provided that it contains all instances of propositional tautologies, all instances of (K) and is closed under the rules (Nec) and (MP);
- a **minimal modal logic** (also called a **classical modal logic**[7]) provided that it contains all instances of propositional tautologies, all instances of (Dual) and is closed under the rules (RE) and (MP);
- a **monotonic modal logic** provided that it contains all instances of propositional tautologies, all instances of (Dual) and all instances of (M) and is closed under the rules (RE) and (MP);
- a **regular modal logic** provided that it contains all instances of propositional tautologies, all instances of (Dual), all instances of (M), all instances of (C) and is closed under the rules (RE) and (MP).

So, **K** is a normal modal logic and **E** is a non-normal modal logic (this follows from Observation 2.23). In fact, **K** is the *smallest* (in terms of the number of theorems) normal modal logic; **E** is the smallest classical modal logic; **EM** is the smallest monotonic modal logic; and **EMC** is the smallest regular modal logic (see Chellas 1980, Chap. 8 for a full discussion).

My first observation about the logics introduced above, is that one can prove a uniform substitution theorem in **E**. Given formulas $\varphi, \psi, \psi' \in \mathcal{L}$, let $\varphi[\psi/\psi']$ be the formula φ but replace *some* occurrences of ψ with ψ' (recall the definition of a substitution from Definition 1.8). For example, suppose that φ is the formula $\Box(\Diamond p \wedge \Box\Box q) \wedge \Box p$; ψ is the formula p; and ψ' is the formula $\Box p$. Then, $\varphi[\psi/\psi']$ can be any of the following

- $\Box(\Diamond p \wedge \Box\Box q) \wedge \Box p$
- $\Box(\Diamond \Box p \wedge \Box\Box q) \wedge \Box p$
- $\Box(\Diamond p \wedge \Box\Box q) \wedge \Box\Box p$
- $\Box(\Diamond \Box p \wedge \Box\Box q) \wedge \Box\Box p$

The uniform substitution theorem states that we can always replace logically equivalent formulas.

Theorem 2.27 (Uniform Substitution) *The following rule can be derived in* **E**:

$$\frac{\psi \leftrightarrow \psi'}{\varphi \leftrightarrow \varphi[\psi/\psi']}$$

---

[7] This is the terminology found in Segerberg (1971) and Chellas (1980). However, this is a somewhat unfortunate name. Starting with Fitch (1948), there is a line of research studying *intuitionistic modal logics* (see, for instance, Artemov and Protopopescu (2016) for an interesting epistemic interpretation of intuitionistic modal logic touching on some of the issues discussed in this book). These are modal logics that extend intuitionistic propositional logics. In this literature, a "classical modal logic" is a modal logics that extends classical propositional logic (as opposed to intuitionistic propositional logic). One may be interested in both normal and non-normal modal logics that extend either classical or intuitionistic propositional logics.

Proof Suppose that ⊢$_E$ $\psi \leftrightarrow \psi'$. We must show that ⊢$_E$ $\varphi \leftrightarrow \varphi[\psi/\psi']$. First of all, note that if φ and ψ are the same formula, then either $\varphi[\psi/\psi']$ is φ (when ψ is not replaced) or $\varphi[\psi/\psi']$ is ψ' (when ψ is replaced). In the first case, $\varphi \leftrightarrow \varphi[\psi/\psi']$ is the formula $\varphi \leftrightarrow \varphi$ and so trivially, ⊢$_E$ $\varphi \leftrightarrow \varphi[\psi/\psi']$. In the second case, $\varphi \leftrightarrow \varphi[\psi/\psi']$ is the formula $\psi \leftrightarrow \psi'$, which is derivable in **E** by assumption. Thus we may assume that φ and ψ are distinct formulas.

The proof is by induction on φ. The base case and Boolean connectives are left as an exercise for the reader. I demonstrate the modal operator. Suppose that φ is $\Box \gamma$ and ⊢$_E$ $\psi \leftrightarrow \varphi'$. The induction hypothesis is ⊢$_E$ $\gamma \leftrightarrow \gamma[\psi/\psi']$. Using the (RE) rule, ⊢$_E$ $\Box\gamma \leftrightarrow \Box(\gamma[\psi/\psi'])$. Note that $\Box(\gamma[\psi/\psi'])$ is the same formula as $\Box\gamma[\psi/\psi']$. Hence, we have ⊢$_E$ $\Box\gamma \leftrightarrow \Box\gamma[\psi/\psi']$. □

The substitution theorem is a fundamental theorem of axiom systems, and will often be implicitly used in the remainder of this book (without reference).

The next observation is that there are alternative characterizations of the logics **EM** and **EN** using the rules (RM) and (Nec), respectively.

Lemma 2.6 *The logic* **EM** *equals the logic* **E** *plus the rule* (RM).

Proof We first show that (RM) can be derived in **EM**.

1. $\varphi \rightarrow \psi$ Assumption
2. $\varphi \leftrightarrow (\varphi \wedge \psi)$ From 1 using propositional logic
3. $\Box\varphi \leftrightarrow \Box(\varphi \wedge \psi)$ From 2 using (RE)
4. $\Box(\varphi \wedge \psi) \rightarrow \Box\varphi \wedge \Box\psi$ Instance of (M)
5. $\Box\varphi \rightarrow \Box\varphi \wedge \Box\psi$ From 3,4 using propositional logic
6. $\Box\varphi \rightarrow \Box\psi$ From 5 using propositional logic

Thus, (RM) is a derived rule of **EM**. The proof that (M) is derivable in the logic **E** plus the rule (RM) is left to the reader. □

Exercise 2.15 Complete the proof of Lemma 2.6.

Lemma 2.7 *The logic* **EN** *equals the logic* **E** *plus the rule* (Nec).

Proof It is easy to see that using (Nec), we can prove $\Box\top$. To see that (Nec) is an admissible rule in the logic **EN**, suppose that ⊢$_{EN}$ φ. We must show that ⊢$_{EN}$ $\Box\varphi$. Using propositional reasoning, since ⊢$_{EN}$ φ, we have ⊢$_{EN}$ $(\top \leftrightarrow \varphi)$. Then, using the rule (RE), ⊢$_{EN}$ $\Box\top \leftrightarrow \Box\varphi$. This means that ⊢$_{EN}$ $\Box\top \rightarrow \Box\varphi$. Since ⊢$_{EN}$ $\Box\top$, by (MP), ⊢$_{EN}$ $\Box\varphi$. □

Given the above Lemmas, it is not hard to see that the logic **EMCN** is equal to the logic **K**. The first step is to show that (K) is derivable in the logic **EMC**:

Lemma 2.8 ⊢$_{EMC}$ (K)

2.3 The Landscape of Non-normal Modal Logics

Proof First of all, since **EMC** contains (M), the rule (RM) is derivable (see Lemma 2.6).

1. $((\varphi \to \psi) \wedge \varphi) \to \psi$ Propositional Tautology
2. $\Box((\varphi \to \psi) \wedge \varphi) \to \Box\psi$ From 1 using (RM)
3. $(\Box(\varphi \to \psi) \wedge \Box\varphi) \to \Box((\varphi \to \psi) \wedge \varphi)$ Instance of (C)
4. $(\Box(\varphi \to \psi) \wedge \Box\varphi) \to \Box\psi$ From 2,3 using propositional logic
5. $\Box(\varphi \to \psi) \to (\Box\varphi \to \Box\psi)$ From 4 using propositional logic

$\Box$

Exercise 2.16 Prove that $\vdash_K$ (M) and $\vdash_K$ (C).

Combining Lemma 2.8 and Exercise 2.16, we have:

Corollary 2.1 *The logic* **EMCN** *equals the logic* **K**.

Notice that both of the deductions that you found to solve the above exercise used the necessitation rule (or the axiom (N)). Using neighborhood structures, it can be shown that all deductions of (M) and (C) in **K** must use the necessitation rule (or the axiom (N) by Lemma 2.7). To show this, we must show that there is a neighborhood frame that validates (K) but not (M) and (C).

Observation 2.28 *The axiom schemes* (M) *and* (C) *are not derivable in* **EK**.

Proof Suppose that $\mathcal{F} = \langle W, N \rangle$ is a neighborhood frame with $W = \{w, v, u, z\}$, and $N(w) = N(v) = N(u) = N(z) = \{\{w, v\}, \{w, u\}\}$. We first show that $\mathcal{F} \models \Box(\varphi \to \psi) \to (\Box\varphi \to \Box\psi)$ (the axiom scheme (K) is valid on $\mathcal{F}$). Suppose that $x \in W$ and $\mathcal{M}$ is any model based on $\mathcal{F}$. Assume that $\mathcal{M}, x \models \Box(\varphi \to \psi)$ and $\mathcal{M}, x \models \Box\varphi$. There are two cases:

1. $[\![\varphi]\!]_\mathcal{M} = \{w, v\}$. Then, $\{u, z\} \subseteq [\![\varphi \to \psi]\!]_\mathcal{M}$. Hence, $\mathcal{M}, x \not\models \Box(\varphi \to \psi)$. This contradicts the first assumption.
2. $[\![\varphi]\!]_\mathcal{M} = \{w, u\}$. Then, $\{v, z\} \subseteq [\![\varphi \to \psi]\!]_\mathcal{M}$. Hence, $\mathcal{M}, x \not\models \Box(\varphi \to \psi)$. This contradicts the first assumption.

Both cases lead to a contradiction. Thus, since it is not possible for $\Box(\varphi \to \psi)$ and $\Box\varphi$ both to be true at a state in a model based on $\mathcal{F}$, the formula $\Box(\varphi \to \psi) \to (\Box\varphi \to \Box\psi)$ is valid on $\mathcal{F}$.

Next, we show that (C) and (M) are not valid on $\mathcal{F}$. The same model works for both formulas. Let $\mathcal{M} = \langle W, N, V \rangle$ be a model based on $\mathcal{F}$ with $V(p) = \{w, v\}$ and $V(q) = \{w, u\}$.

- We have $[\![p]\!]_\mathcal{M} = \{w, v\}$ and $[\![q]\!]_\mathcal{M} = \{w, u\}$ and $[\![p \wedge q]\!]_\mathcal{M} = \{w\}$. $\mathcal{M}, w \models \Box p \wedge \Box q$, but $\mathcal{M}, w \not\models \Box(p \wedge q)$.
- We have $[\![(p \vee q) \wedge p]\!]_\mathcal{M} = \{w, v\}$ and $[\![p \vee q]\!]_\mathcal{M} = \{w, v, u\}$. Then, $\mathcal{M}, w \models \Box((p \vee q) \wedge p)$; however, $\mathcal{M}, w \not\models \Box(p \vee q)$. Thus, $\Box(\varphi \wedge \psi) \to \Box\varphi \wedge \Box\psi$ is not valid. $\Box$

Exercise 2.17 1. Prove that $\vdash_{\mathbf{E}} \Diamond \top \leftrightarrow \neg \Box \bot$ and $\vdash_{\mathbf{E}} \Box \top \leftrightarrow \neg \Diamond \bot$.
2. Prove that the following are theorems of any monotonic modal logic:
   a. $\Box \psi \to \Box(\varphi \to \psi)$
   b. $\Box \neg \varphi \to \Box(\varphi \to \psi)$
   c. $\Diamond(\varphi \to \psi) \vee \Box(\psi \to \varphi)$
   d. $\Diamond(\varphi \wedge \psi) \to \Diamond \varphi \wedge \Diamond \psi$
3. Prove that the rule $\frac{\varphi \to \psi}{\Diamond \varphi \to \Diamond \psi}$ is *admissible* in any monotonic modal logic. (A rule is *admissible* in a logic if adding it does not change the set of theorems.)
4. Prove that the following are derivable in any regular modal logic:
   a. $\Box(\varphi \to \psi) \to (\Diamond \varphi \to \Diamond \psi)$
   b. $\Box(\varphi \leftrightarrow \psi) \to (\Box \varphi \leftrightarrow \Box \psi)$
   c. $(\Box \varphi \wedge \Diamond \psi) \to \Diamond(\varphi \wedge \psi)$
5. Prove that $(\Box \varphi \to \Diamond \varphi) \to \Diamond \top$ is derivable in a monotonic modal logic.
6. Find a monotonic neighborhood model $\mathcal{M}$ with a state w such that $\mathcal{M}, w \not\models \Diamond \top \to (\Box p \to \Diamond p)$. (This shows that $\not\vdash_{\mathbf{EM}} \Diamond \top \to (\Box \varphi \to \Diamond \varphi)$.)
7. Prove that $\Diamond \top \leftrightarrow (\Box \varphi \to \Diamond \varphi)$ is a theorem of every regular modal logic.

2.3.1 A Non-normal Extension of K

I conclude this section with some general comments about the definition of (normal) modal logics. By definition, a *logic* is a set of formulas (typically, instances of some collection of axiom schemes) that is closed under some inference rules. Thus, the statement "the logic $\mathbf{L}_1$ is contained in the logic $\mathbf{L}_2$" means that all the formulas in $\mathbf{L}_1$ are in $\mathbf{L}_2$ (i.e., $\mathbf{L}_1 \subseteq \mathbf{L}_2$). From this point of view, it is a direct consequence of Definition 2.26 that if $\mathbf{L}$ is a normal modal logic, then $\mathbf{L}$ contains $\mathbf{K}$. It is, perhaps, surprising to note that it is not the case that every logic that contains $\mathbf{K}$ is normal. An example of such a logic was provided early on by McKinsey and Tarski (1944). I will present this logic below following the discussion in Segerberg (1971, pp. 171, 172).[8]

I start by defining a well-known normal modal logic. Let **S4** be the smallest set of formulas from $\mathcal{L}(\mathsf{At})$ that is closed under (MP) and the necessitation rule (Nec), and that contains all propositional tautologies, all instances of (K), and all instances of the following axiom schemes:

$$(\mathsf{T}) \quad \Box \varphi \to \varphi \quad \text{and} \quad (4) \quad \Box \varphi \to \Box \Box \varphi.$$

The logics that interest us in this section contain the so-called *McKinsey axiom*:

$$(\mathsf{McK}) \quad \Box \Diamond \varphi \to \Diamond \Box \varphi.$$

Consider the following two logics:

---

[8]This is a small digression that can easily be skipped by readers not already familiar with relational semantics for modal logic (see Appendix A for the relevant definitions).

2.3 The Landscape of Non-normal Modal Logics

1. **S4McK** is the smallest set of formulas that contains all instances of (Dual), (K), (T), (4) and (McK) and is closed under (MP) and (Nec).
2. **L** is the smallest set of formulas that contains all of **S4** and all instances of the axiom scheme (McK).

Clearly, **S4McK** is a normal modal logic. The logic **L** contains **S4** (and, so, also contains the logic **K**). The question is: are **L** and **S4McK** the same logic? I will argue that the two logics are distinct. In particular, $\Box(\Box\Diamond p \to \Diamond\Box p) \notin \mathbf{L}$ (of course, $\Box(\Box\Diamond p \to \Diamond\Box p) \in \mathbf{S4McK}$). Thus, **L** is a non-normal modal logic that contains **K**. To see that $\Box(\Box\Diamond p \to \Diamond\Box p) \notin \mathbf{L}$, consider the relational frame (Definition A.1) $\mathfrak{F} = \langle W, R \rangle$, where $W = \{w_1, w_2, w_3, w_4\}$, and R is the smallest reflexive relation containing

$$\{(w_1, w_2), (w_1, w_3), (w_1, w_4), (w_3, w_4), (w_4, w_3)\}.$$

This frame can be depicted as follows:

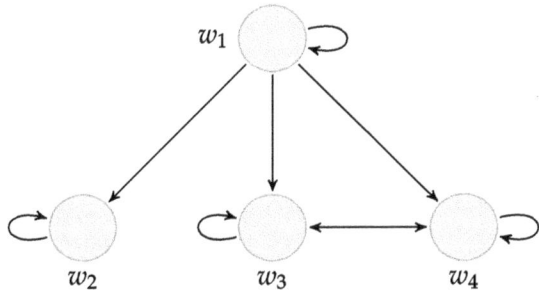

Consider the set of formulas that are true at w_1 in any relational model based on the above frame:

$$\mathbf{L}_{w_1} = \{\varphi \mid \mathfrak{M}, w_1 \models \varphi \text{ where } \mathfrak{M} \text{ is a relational model based on } \mathfrak{F}\}.$$

The following two Claims establish the fact that $\Box(\Box\Diamond p \to \Diamond\Box p) \notin \mathbf{L}$.

Claim $\mathbf{L} \subseteq \mathbf{L}_{w_1}$

Proof We must show that $\mathbf{S4} \subseteq \mathbf{L}_{w_1}$, and that every instance of (McK) is contained in $\mathbf{L}_{w_1}$. The fact that $\mathbf{S4} \subseteq \mathbf{L}_{w_1}$ follows from the fact that the relation in $\mathfrak{F}$ is reflexive and transitive, and from the well-known result that **S4** is sound and complete with respect to the class of frames that are reflexive and transitive (Blackburn et al. 2001, Theorem 4.29). To see that every instance of (McK) is contained in $\mathbf{L}_{w_1}$, suppose that $\mathfrak{M}$ is a relational model based on $\mathfrak{F}$ and $\mathfrak{M}, w_1 \models \Box\Diamond\varphi$. Then, for all v, if $w_1 R v$, then $\mathfrak{M}, v \models \Diamond\varphi$. In particular, since $w_1 R w_2$, $\mathfrak{M}, w_2 \models \Diamond\varphi$. Thus, since w_2 is the only world accessible from w_2, we must have $\mathfrak{M}, w_2 \models \varphi$. Furthermore, $\mathfrak{M}, w_2 \models \Box\varphi$ (this again follows from the fact that w_2 is the only world accessible from w_2). But this means that $\mathfrak{M}, w_1 \models \Diamond\Box\varphi$. Thus, $\mathfrak{M}, w_1 \models \Box\Diamond\varphi \to \Diamond\Box\varphi$. □

Claim $\Box(\Box\Diamond p \to \Diamond\Box p) \notin \mathbf{L}_{w_1}$

Proof Suppose that $\mathfrak{M} = \langle W, R, V \rangle$ is a relational model based on $\mathfrak{F}$ with the valuation $V(p) = \{w_3\}$. Then, as the reader is invited to verify, $\mathfrak{M}, w_4 \models \Box\Diamond p$, but $\mathfrak{M}, w_4 \not\models \Diamond\Box p$. This means that $\mathfrak{M}, w_4 \not\models \Box\Diamond p \to \Diamond\Box p$, and so $\mathfrak{M}, w_1 \not\models \Box(\Box\Diamond p \to \Diamond\Box p)$. Thus, $\Box(\Box\Diamond p \to \Diamond\Box p) \notin \mathbf{L}_{w_1}$, as desired. □

Remark 2.29 (*Another Example*) The example of a non-normal modal logic containing **K** presented in this section is somewhat artificial. A better motivated example of a such a logic is Solovay's provability logic **S** (see Japaridze and de Jongh 1998, for a discussion). The normal modal logic **GL** is defined by adding the axiom scheme $\Box(\Box\varphi \to \varphi) \to \Box\varphi$ to **K**. It is well-known that $\vdash_{\mathbf{GL}} \Box\varphi \to \Box\Box\varphi$. Solovay's logic **S** is defined as follows: $\varphi \in \mathbf{S}$ iff $\{\Box\psi_1 \to \psi_1, \ldots, \Box\psi_k \to \psi_k\} \vdash_{\mathbf{GL}} \varphi$, for some $\psi_1, \ldots, \psi_k$. It can be shown that **S** contains **K4**, but is not closed under the rule of necessitation (Japaridze and de Jongh 1998, Sect. 2). Thus, **S** is another example of a non-normal modal logic containing **K**.

2.3.2 Completeness

In this section, I show how to adapt the standard approach for proving completeness of modal logics to prove completeness of non-normal modal logics with respect to neighborhood semantics. I assume that the reader is familiar with basic soundness and completeness results in modal logic (with respect to relational frames). See Blackburn et al. (2001, Chap. 4) for an overview. I start by reviewing some basic terminology.

2.3.2.1 Preliminaries

Recall the definition of a deduction for modal logic from (Definition 2.25 and the subsequent discussion in Sect. 2.3).

Definition 2.30 (*Deduction from assumptions*) Suppose that Γ is a set of formulas from $\mathcal{L}$; **L** is a modal logic; and $\varphi \in \mathcal{L}$. We write $\Gamma \vdash_{\mathbf{L}} \varphi$ if there is a finitely many formulas $\alpha_1, \ldots, \alpha_n \in \Gamma$ such that $\vdash_{\mathbf{L}} (\alpha_1 \wedge \cdots \wedge \alpha_n) \to \varphi$.[9]

Suppose that F is a collection of neighborhood frames. A formula $\varphi \in \mathcal{L}$ is **valid in** F, or **F-valid**, denoted $\models_{\mathsf{F}} \varphi$, when for each $\mathcal{F} \in \mathsf{F}$, $\mathcal{F} \models \varphi$ (Definition 1.13). Given a class of frame F, let $\mathbf{L}(\mathsf{F}) = \{\varphi \mid \text{for all } \mathcal{F} \in \mathsf{F}, \mathcal{F} \models \varphi\}$ denote the set of formulas that are F-valid.

---

[9]I am using the definition of a deduction from assumptions found in Goldblatt (1992a, p. 17) and Blackburn et al. (2001, p. 36). See Hakli and Negri (2011) for a discussion of the issues surrounding this definition related to the deduction theorem and the proper use of inference rules.

2.3 The Landscape of Non-normal Modal Logics

Definition 2.31 (*Soundness*) A logic **L** is **sound** with respect to F, provided that $\mathbf{L} \subseteq \mathbf{L}(\mathsf{F})$. That is, for each $\varphi \in \mathcal{L}$, if $\vdash_\mathbf{L} \varphi$, then φ is valid in F.

The main goal is to show that there is a semantic consequence relation between sets of formulas and formulas that is equivalent to the deduction relation.

Definition 2.32 (*Semantic Consequence*) Suppose that Γ is a set of formulas and F is a class of neighborhood frames. A formula $\varphi \in \mathcal{L}$ is a **semantic consequence** with respect to F of Γ, denoted $\Gamma \models_\mathsf{F} \varphi$, provided for each model $\mathcal{M} = \langle W, N, V \rangle$ based on a frame from F (i.e., $\langle W, N \rangle \in \mathsf{F}$), for each $w \in W$, if $\mathcal{M}, w \models \Gamma$, then $\mathcal{M}, w \models \varphi$.

Remark 2.33 (*Local and Global Consequence*) The above definition of a semantic consequence is a **local consequence** relation. It is important to distinguish this from a **global consequence**: For a class of frames F, let $\Gamma \models_\mathsf{F}^g \varphi$ provided that for each $\mathcal{F} \in \mathsf{F}$, if $\mathcal{F} \models \Gamma$, then $\mathcal{F} \models \varphi$. These two notions of semantic consequence are not equivalent. For instance, suppose that F is the class of frames that contain the unit (i.e., for all $\langle W, N \rangle \in \mathsf{F}$, for all $w \in W$, $W \in N(w)$); then, $\{p\} \models_\mathsf{F}^g \Box p$ but $\{p\} \not\models_\mathsf{F} \Box p$. See Blackburn et al. (2001, Sect. 1.5) and Hakli and Negri (2011) for further discussion.

Definition 2.34 (*Strong Completeness*) A logic **L** is **strongly complete** with respect to a class of frames F, when, for each $\Gamma \subseteq \mathcal{L}$, $\Gamma \models_\mathsf{F} \varphi$ implies $\Gamma \vdash_\mathbf{L} \varphi$.

Remark 2.35 (*Weak Completeness*) A special case of the above definition is when $\Gamma = \emptyset$. A logic **L** is **weakly complete** with respect to a class of frames F, if $\models_\mathsf{F} \varphi$ implies $\vdash_\mathbf{L} \varphi$. Obviously, if **L** is strongly complete, then **L** is weakly complete. However, the converse is not true. There are modal logics that are weakly complete but not strongly complete. An example of such a normal modal logic can be found in the Appendix (Observation A.19). See, also, Sect. 3.3 for other examples of logics that are weakly complete but no strongly complete (cf., also, Blackburn et al. (2001), Sect. 4.8).

Let **L** be any modal logic extending **E**. A set of formulas Γ is said to be **L-inconsistent** if $\Gamma \vdash_\mathbf{L} \bot$, and **L-consistent** if it is not inconsistent.

Definition 2.36 (*Maximally Consistent Set*) A set of formulas Γ is called **maximally consistent** provided that Γ is consistent and for all formulas $\varphi \in \mathcal{L}$, either $\varphi \in \Gamma$ or $\neg \varphi \in \Gamma$.

Let $M_\mathbf{L}$ be the set of **L**-maximally consistent sets of formulas. Given a formula $\varphi \in \mathcal{L}$, let $|\varphi|_\mathbf{L}$ be the **proof set** of φ in **L**. Formally, $|\varphi|_\mathbf{L} = \{\Delta \mid \Delta \in M_\mathbf{L} \text{ and } \varphi \in \Delta\}$. The first observation is that proof sets share a number of properties in common with truth sets.

Lemma 2.9 *Let* **L** *be a logic and* $\varphi, \psi \in \mathcal{L}$. *Then,*

1. $|\varphi \wedge \psi|_{\mathbf{L}} = |\varphi|_{\mathbf{L}} \cap |\psi|_{\mathbf{L}}$.
2. $|\neg \varphi|_{\mathbf{L}} = M_{\mathbf{L}} - |\varphi|_{\mathbf{L}}$.
3. $|\varphi \vee \psi|_{\mathbf{L}} = |\varphi|_{\mathbf{L}} \cup |\psi|_{\mathbf{L}}$.
4. $|\varphi|_{\mathbf{L}} \subseteq |\psi|_{\mathbf{L}}$ *iff* $\vdash_{\mathbf{L}} \varphi \to \psi$.
5. $|\varphi|_{\mathbf{L}} = |\psi|_{\mathbf{L}}$ *iff* $\vdash_{\mathbf{L}} \varphi \leftrightarrow \psi$.
6. *For any maximally* **L**-*consistent set* Γ, *if* $\varphi \in \Gamma$ *and* $\varphi \to \psi \in \Gamma$, *then* $\psi \in \Gamma$.
7. *For any maximally* **L**-*consistent set* Γ, *If* $\vdash_{\mathbf{L}} \varphi$, *then* $\varphi \in \Gamma$.

Exercise 2.18 Prove Lemma 2.9.

Another standard result is *Lindenbaum's Lemma*. I leave the proof of Lindenbaum's Lemma as an exercise.[10]

Lemma 2.10 (Lindenbaum's Lemma) *For any* **L**-*consistent set of formulas* Γ, *there exists a maximally* **L**-*consistent set* Γ' *such that* $\Gamma \subseteq \Gamma'$.

Exercise 2.19 Prove Lindenbaum's Lemma.

The following useful fact about proof sets demonstrates how Lindenbaum's Lemma can be used.

Lemma 2.11 *For each* $\varphi \in \mathcal{L}$, $\psi \in \bigcap |\varphi|_{\mathbf{L}}$ *iff* $\vdash_{\mathbf{L}} \varphi \to \psi$.

Proof Suppose that $\vdash_{\mathbf{L}} \varphi \to \psi$. Then, for each maximally consistent set Γ, $\varphi \to \psi \in \Gamma$. Hence, since for each $\Gamma \in |\varphi|_{\mathbf{L}}$, $\varphi \in \Gamma$, we have $\psi \in \Gamma$. Thus, $\psi \in \bigcap |\varphi|_{\mathbf{L}}$.

Suppose that $\psi \in \bigcap |\varphi|_{\mathbf{L}}$, but it is not the case that $\vdash_{\mathbf{L}} \varphi \to \psi$. Then, $\neg(\varphi \to \psi)$ is **L**-consistent. Using Lindenbaum's Lemma, there is a maximally consistent set Γ such that $\neg(\varphi \to \psi) \in \Gamma$. Thus, $\varphi, \neg\psi \in \Gamma$. Since $\varphi \in \Gamma$, $\Gamma \in |\varphi|_{\mathbf{L}}$. But then, $\neg\psi \in \Gamma$ contradicts the fact that $\psi \in \bigcap |\varphi|_{\mathbf{L}}$. □

A straightforward corollary of this Lemma is the following useful fact:

Corollary 2.2 *If* $\varphi \in \Gamma$ *for all maximally* **L**-*consistent sets* Γ, *then* $\vdash_{\mathbf{L}} \varphi$.

2.3.2.2 Canonical Models

Suppose that $\mathcal{M} = \langle W, N, V \rangle$ is a neighborhood model and $X \subseteq W$ any subset. A set $X \subseteq W$ is **definable** (with respect to a modal language $\mathcal{L}$) provided that there is a formula $\varphi \in \mathcal{L}$ such that $[\![\varphi]\!]_{\mathcal{M}} = X$. Let $\mathcal{D}_{\mathcal{M}}$ be the set of all sets that are definable in $\mathcal{M}$. Note that since there are only countably many formulas in $\mathcal{L}$, the set $\mathcal{D}_{\mathcal{M}}$ is countable (or finite if W is finite). Thus, if $\wp(W)$ is uncountable, then $\mathcal{D}_{\mathcal{M}} \neq \wp(W)$. A subset $X \subseteq M_{\mathbf{L}}$ is called a **proof set** provided that there is some formula $\varphi \in \mathcal{L}$ such that $X = |\varphi|_{\mathbf{L}}$. Again, since the modal language $\mathcal{L}$ is countable,

---

[10] The proof is provided in the solution manual. Consult Chellas (1980) and Blackburn et al. (2001) for a discussion of this proof and a more complete discussion of maximally consistent sets.

2.3 The Landscape of Non-normal Modal Logics

there are only countably many proof sets. However, if At is countable, then $M_\mathbf{L}$ is uncountable; and so, there are uncountably many subsets of $M_\mathbf{L}$. Thus, it is not the case that every subset of $M_\mathbf{L}$ is a proof set.

As usual, the states in a canonical model will be maximally consistent sets—i.e., elements of $M_\mathbf{L}$. A function $N_\mathbf{L} : M_\mathbf{L} \to \wp(\wp(M_\mathbf{L}))$ is a **canonical neighborhood function** provided that for all $\varphi \in \mathcal{L}$:

$$|\varphi|_\mathbf{L} \in N_\mathbf{L}(\Gamma) \text{ iff } \Box\varphi \in \Gamma.$$

So, for each maximally consistent set Γ, $N_\mathbf{L}(\Gamma)$ contains at least all the proof sets of the necessary formulas from Γ. The first question is: *Do any such functions actually exist*? That is, is it even possible to define a function satisfying the above condition? A problem would arise if there were proof sets $|\varphi|_\mathbf{L}$ and $|\psi|_\mathbf{L}$ such that $|\varphi|_\mathbf{L} = |\psi|_\mathbf{L}$ and a maximally consistent set with $\Box\varphi \in \Gamma$ but $\Box\psi \notin \Gamma$ (and, hence, $\neg\Box\psi \in \Gamma$). If such a situation were possible, then it would be impossible to satisfy the above condition. Fortunately, this problematic situation is ruled out in any logic containing the rule RE.

Lemma 2.12 *Suppose that* $\mathbf{L}$ *is a logic that contains the RE rule and that* $N_\mathbf{L} : M_\mathbf{L} \to \wp(\wp(M_\mathbf{L}))$ *is a function such that for each* $\Gamma \in M_\mathbf{L}$, $|\varphi|_\mathbf{L} \in N_\mathbf{L}(\Gamma)$ *iff* $\Box\varphi \in \Gamma$. *Then, if* $|\varphi|_\mathbf{L} \in N_\mathbf{L}(\Gamma)$ *and* $|\varphi|_\mathbf{L} = |\psi|_\mathbf{L}$, *then* $\Box\psi \in \Gamma$.

Proof Let φ and ψ be two formulas such that $|\varphi|_\mathbf{L} = |\psi|_\mathbf{L}$, $\Box\varphi \in \Gamma$, and $|\varphi|_\mathbf{L} \in N_\mathbf{L}(\Gamma)$. Since $|\varphi|_\mathbf{L} \in N_\mathbf{L}(\Gamma)$, $\Box\varphi \in \Gamma$. Also, by Lemma 2.9, since $|\varphi|_\mathbf{L} = |\psi|_\mathbf{L}$, $\vdash_\mathbf{L} \varphi \leftrightarrow \psi$. Using the RE rule, $\vdash_\mathbf{L} \Box\varphi \leftrightarrow \Box\psi$. Hence, $\Box\varphi \leftrightarrow \Box\psi \in \Gamma$. Therefore, $\Box\psi \in \Gamma$. □

The **canonical valuation**, $V_\mathbf{L} : \text{At} \to \wp(M_\mathbf{L})$, is defined as follows. For each $p \in \text{At}$, let $V_\mathbf{L}(p) = |p|_\mathbf{L} = \{\Gamma \mid \Gamma \in M_\mathbf{L} \text{ and } p \in \Gamma\}$. Putting everything together gives us the following definition:

Definition 2.37 (*Canonical Neighborhood Model*) Suppose that $\mathcal{M} = \langle W, N, V \rangle$ is a neighborhood model. Then, $\mathcal{M}$ is **canonical for L** provided that:

1. $W = M_\mathbf{L}$;
2. for each $\Gamma \in W$ and each formula $\varphi \in \mathcal{L}$, $|\varphi|_\mathbf{L} \in N(\Gamma)$ iff $\Box\varphi \in \Gamma$; and
3. $V = V_\mathbf{L}$.

For example, let $\mathcal{M}_\mathbf{L}^{min} = \langle M_\mathbf{L}, N_\mathbf{L}^{min}, V_\mathbf{L} \rangle$, where, for each $\Gamma \in M_\mathbf{L}$, $N_\mathbf{L}^{min}(\Gamma) = \{|\varphi|_\mathbf{L} \mid \Box\varphi \in \Gamma\}$. The model $\mathcal{M}_\mathbf{L}^{min}$ is easily seen to be canonical for $\mathbf{L}$. Furthermore, it is the minimal canonical for $\mathbf{L}$ in the sense that for each Γ, $N_\mathbf{L}^{min}(\Gamma)$ is the smallest set satisfying property 2 from Definition 2.37. Let $\mathcal{P}_\mathbf{L}$ be the set of all proof sets of $\mathbf{L}$ (i.e., $\mathcal{P}_\mathbf{L} = \{|\varphi|_\mathbf{L} \mid \varphi \in \mathcal{L}\}$). The largest canonical for $\mathbf{L}$ is the model $\mathcal{M}_\mathbf{L}^{max} = \langle M_\mathbf{L}, N_\mathbf{L}^{max}, V_\mathbf{L} \rangle$ with for each $\Gamma \in M_\mathbf{L}$, $N_\mathbf{L}^{max}(\Gamma) = N_\mathbf{L}^{min}(\Gamma) \cup \{X \mid X \subseteq M_\mathbf{L}, X \notin \mathcal{P}_\mathbf{L}\}$.

Lemma 2.13 (Truth Lemma) *For any consistent logic* **L** *and any consistent formula* φ, *if* $\mathcal{M}$ *is canonical for* **L**, *then*

$$[\![\varphi]\!]_\mathcal{M} = |\varphi|_\mathbf{L}.$$

Proof Suppose that $\mathcal{M} = \langle W, N, V \rangle$ is a canonical model for **L**. The proof is by induction on the structure of $\varphi \in \mathcal{L}$. The base case and the cases for the Boolean connectives are as usual and left for the reader. I give the details only for the modal case. The induction hypothesis is that $[\![\varphi]\!]_\mathcal{M} = |\varphi|_\mathbf{L}$. We must show that $[\![\Box\varphi]\!]_\mathcal{M} = |\Box\varphi|_\mathbf{L}$.

$$\begin{aligned}
\Gamma \in [\![\Box\varphi]\!]_\mathcal{M} &\text{ iff } [\![\varphi]\!]_\mathcal{M} \in N(\Gamma) &&\text{(Definition of truth)} \\
&\text{ iff } |\varphi|_\mathbf{L} \in N(\Gamma) &&\text{(induction hypothesis)} \\
&\text{ iff } \Box\varphi \in \Gamma &&\text{(item 2 of Definition 2.37)} \\
&\text{ iff } \Gamma \in |\Box\varphi|_\mathbf{L} &&\text{(definition of proof sets)}
\end{aligned}$$

Thus, $[\![\Box\varphi]\!]_\mathcal{M} = |\Box\varphi|_\mathbf{L}$, as desired. □

2.3.2.3 Applications

Theorem 2.38 *The logic* **E** *is sound and strongly complete with respect to the class of all neighborhood frames.*

Proof Soundness is straightforward (and, in fact, already shown in earlier exercises). As for strong completeness, I will show that every consistent set of formulas can be satisfied in some model. Before proving this, I show that this fact implies strong completeness. The proof is by contraposition. Suppose that it is not the case that $\Gamma \vdash_\mathbf{L} \varphi$. Then, $\Gamma \cup \{\neg\varphi\}$ is consistent. (If $\Gamma \cup \{\neg\varphi\} \vdash_\mathbf{L} \bot$, then, by propositional reasoning, $\Gamma \vdash_\mathbf{L} \neg\varphi \to \bot$. Hence, $\Gamma \vdash_\mathbf{L} \varphi$.) If $\Gamma \cup \{\neg\varphi\}$ is jointly true at some state in a model, then Γ cannot semantically entail φ. Thus, if $\Gamma \not\vdash_\mathbf{L} \varphi$, then $\Gamma \not\models_\mathsf{F} \varphi$ (where F is the class of all neighborhood frames).

Let Γ be a consistent set of formulas. By Lindenbaum's Lemma, there is a maximally consistent set Γ' such that $\Gamma \subseteq \Gamma'$. Consider the minimal canonical model $\mathcal{M}_\mathbf{E}^{min}$. By the Truth Lemma (Lemma 2.13), $\mathcal{M}_\mathbf{L}^{min}, \Gamma' \models \Gamma'$. Thus, Γ is satisfiable at a state in the minimal canonical model, as desired. □

Notice that in the above proof, the choice to use the *minimal* canonical model for **E** was somewhat arbitrary. It is easy to see that the proof would go through if I had used $\mathcal{M}_\mathbf{E}^{max}$ instead of $\mathcal{M}_\mathbf{E}^{min}$. Indeed, *any* canonical model for **E** could have been used in the above proof. The fact that there is a choice of canonical models will be useful when proving completeness for logics extending **E**. The strategy for proving strong completeness for other non-normal modal logics discussed in Sect. 2.3 is similar to the strategy for proving strong completeness of some well-known normal modal logics, such as **S4** or **S5**. Given the above definition of a canonical model and truth lemma, all that remains is to show that a the frame of a particular canonical

2.3 The Landscape of Non-normal Modal Logics

model belongs to the class of frames under consideration. This argument is called *completeness-via-canonicity* in Blackburn et al. (2001). For instance, consider the logic **EC**. It is not hard to see that **C** is sound for the class of neighborhood frames that are closed under intersections (cf. Lemma 2.20). I now show that **EC** is sound and strongly complete with respect to neighborhood frames that are closed under intersections. The first step is to show that C is **canonical**[11] for this property.

Lemma 2.14 *If* **L** *contains all instances of C, then* $N_{\mathbf{L}}^{min}$ *is closed under finite intersections.*

Proof Suppose that **L** contains all instances of C. We must show that for all $\Gamma \in M_{\mathbf{L}}$, $N_{\mathbf{L}}^{min}(\Gamma)$ is closed under intersections. Suppose that $X, Y \in N_{\mathbf{L}}^{min}(\Gamma)$. By the definition of $N_{\mathbf{L}}^{min}$, $X = |\varphi|_{\mathbf{L}}$ and $Y = |\psi|_{\mathbf{L}}$ with $\Box \varphi \in \Gamma$ and $\Box \psi \in \Gamma$. Hence, $\Box \varphi \wedge \Box \psi \in \Gamma$; and so, using C, $\Box(\varphi \wedge \psi) \in \Gamma$. Thus, $|\varphi \wedge \psi|_{\mathbf{L}} \in N_{\mathbf{L}}^{min}(\Gamma)$. Therefore, $X \cap Y = |\varphi|_{\mathbf{L}} \cap |\psi|_{\mathbf{L}} = |\varphi \wedge \psi|_{\mathbf{L}} \in N_{\mathbf{L}}^{min}(\Gamma)$, as desired. □

Given the above proof, strong completeness is straightforward.

Theorem 2.39 *The logic* **EC** *is sound and strongly complete with respect to the class of neighborhood frames that are closed under intersections.*

Proof The proof proceeds as in Theorem 2.38, using Lemma 2.14 to argue that the canonical frame for **EC** is closed under intersections. □

Exercise 2.20 Prove that **EN** is sound and strongly complete with respect to neighborhood frames that contain the unit.

The proof that **EM** is strongly complete with respect to neighborhood frames that are closed under supersets is not as straightforward. Here, we need to make use of the fact that there are a number of different canonical models. The main difficulty is that $N_{\mathbf{EM}}^{min}$ is not closed under supersets.

Observation 2.40 *There is a maximally consistent set* Γ *such that* $N_{\mathbf{EM}}^{min}(\Gamma)$ *is not closed under supersets.*

Proof Let p be a propositional variable and let Γ be a maximally consistent set such that $\Box p \in \Gamma$ (such a set exists by Lindenbaum's Lemma since $\Box p$ is consistent). Then $|p|_{\mathbf{EM}} \in N_{\mathbf{EM}}^{min}(\Gamma)$. Let Y be any non-proof set that extends $|p|_{\mathbf{EM}}$ (i.e., $|p|_{\mathbf{EM}} \subsetneq Y$). To see that such a set exists, let Y' be any non-proof set such that $Y' \not\subseteq |p|_{\mathbf{EM}}$ (such a set exists since there are uncountably many subsets of $M_{\mathbf{EM}}$ but only countably many proof sets, and p is not a theorem of **EM**). Then, $Y = Y' \cup |p|_{\mathbf{EM}}$ is not a proof set since, if $Y = |\psi|_{\mathbf{EM}}$ for some formula ψ, then $Y' = |\psi \wedge \neg p|_{\mathbf{EM}}$ (why?), which contradicts the fact that Y' is not a proof set. Clearly, $Y \notin N_{\mathbf{EM}}^{min}(\Gamma)$ (why?). Then, we have $X = |p|_{\mathbf{EC}} \in N_{\mathbf{EM}}^{min}(\Gamma)$, $X \subseteq Y$, but $Y \notin N_{\mathbf{EM}}^{min}(\Gamma)$. □

---

[11] See Blackburn et al. (2001), Chap. 4, for an extended discussion of canonical properties for relational models.

However, this difficulty can be easily overcome by choosing a different, better-behaved, canonical model. Recall from Sect. 1.1, that if $\mathcal{F}$ is any collection of subsets of W, then $\mathcal{F}^{mon} = \{X \mid \text{there is a } Y \in \mathcal{F} \text{ such that } Y \subseteq X\}$. Given any model $\mathcal{M} = \langle W, N, V \rangle$, let the **supplementation** of $\mathcal{M}$, denoted $\mathcal{M}^{mon}$, be the model $\langle W, N^{mon}, V \rangle$, where for each $w \in W$, $N^{mon}(w) = (N(w))^{mon}$. The key argument is that the supplementation of the minimal canonical model is canonical for **EM**.

Lemma 2.15 *Suppose that* $\mathcal{M} = (\mathcal{M}_{\mathbf{EM}}^{min})^{mon}$. *Then,* $\mathcal{M}$ *is canonical for* **EM**.

Proof Suppose that $\mathcal{M} = \langle W, N, V \rangle$, where $W = M_{\mathbf{EM}}$ and for each $\Gamma \in W$, $N(\Gamma) = (N_{\mathbf{EM}}^{min}(\Gamma))^{mon}$, and $V = V_{\mathbf{EM}}$. Let $\Gamma \in W$. We must show that for each formula $\varphi \in \mathcal{L}$,
$$|\varphi|_{\mathbf{EM}} \in N(\Gamma) \text{ iff } \Box\varphi \in \Gamma.$$

The right-to-left direction is trivial since for all Γ, $N_{\mathbf{EM}}^{min}(\Gamma) \subseteq N(\Gamma)$. Suppose that $|\varphi|_{\mathbf{EM}} \in N(\Gamma) = (N_{\mathbf{EM}}^{min}(\Gamma))^{mon}$. Then, there is some proof set $|\psi|_{\mathbf{EM}} \in N_{\mathbf{EM}}^{min}(\Gamma)$ such that $|\psi|_{\mathbf{EM}} \subseteq |\varphi|_{\mathbf{EM}}$. Since $|\psi|_{\mathbf{EM}} \in N_{\mathbf{EM}}^{min}(\Gamma)$, we have $\Box\psi \in \Gamma$. Furthermore, since $|\psi|_{\mathbf{EM}} \subseteq |\varphi|_{\mathbf{EM}}$, by Lemma 2.9, $\vdash_{\mathbf{EM}} \psi \to \varphi$. Using the rule RM (which is admissible in **EM**), $\vdash_{\mathbf{EM}} \Box\psi \to \Box\varphi$. Thus, $\Box\psi \to \Box\varphi \in \Gamma$. Therefore, $\Box\varphi \in \Gamma$, as desired. $\square$

Theorem 2.41 *The logic* **EM** *is sound and strongly complete with respect to the class of monotonic frames.*

Proof Left as an exercise for the reader. $\square$

Combining the proofs of Theorems 2.39 and 2.41 with Exercise 2.20 gives a characterization of the smallest normal modal logic **K**.

Theorem 2.42 *The logic* **K** *is sound and strongly complete with respect to the class of filters.*

As I explained in Sect. 1.1, not all filters are augmented. Since **K** is sound and complete with respect to the class of all relational frames (cf. Appendix A), and each relational frame corresponds to an augmented neighborhood frame (cf. Sect. 2.2.1), there is another characterization of **K**:

Exercise 2.21 Prove that **K** is sound and strongly complete with respect to the class of augmented frames.

Exercise 2.22 Prove that **S4** (see Sect. 2.3.1 for a definition) is sound and strongly complete with respect to the class of **S4** neighborhood frames (Definition 1.28).

2.3.3 Incompleteness and General Frames

In Sect. 2.2.1, we saw that the class of relational frames is modally equivalent to the class of augmented neighborhood frames. This means that if a modal logic is

2.3 The Landscape of Non-normal Modal Logics

complete with respect to a class of relational frames, then it must be complete with respect to the corresponding class of neighborhood frames (this follows from the fact that every relational model can be turned into a modally equivalent neighborhood model). Recall that a logic **L** is **neighborhood complete** (resp. **Kripke complete**) provided that there is a class of neighborhood frames F (resp. relational frames) such that $\mathbf{L} = \mathbf{L}(\mathsf{F}) = \{\varphi \in \mathcal{L} \mid \mathcal{F} \models \varphi \text{ for all } \mathcal{F} \in \mathsf{F}\}$. Otherwise, the logic is said to be **neighborhood incomplete** (resp. **Kripke incomplete**).

It is well known that there are Kripke incomplete modal logics—i.e., modal logics that are not the logic of any class for relational frames. Thomason (1972) provided the first consistent modal logic that is incomplete with respect to relational frames (i.e., a Kripke incomplete modal logic). See Fine (1974), Thomason (1974), van Benthem (1978), Shehtman (1977), Boolos and Sambin (1985), and Benton (2002) for other examples of Kripke incomplete modal logics. Gerson (1975b) proved that the Kripke incomplete modal logics from Thomason (1974) and Fine (1974) are also incomplete with respect to neighborhood frames (cf. also Shehtman (1980) for another example a logic incomplete with respect to neighborhood frames). Now, if a logic is non-normal, then, since there are no relational frames validating all the formulas in the logic, the logic must be Kripke incomplete. However, the situation is much more interesting. There are consistent normal modal logics that are complete with respect to a class of neighborhood frames but not complete with respect to any class of relational frames. Gerson showed this for a logic extending **S4** (Gerson 1975a) and a logic extending **T** (Gerson 1976). Other examples of modal logics that are complete with respect to neighborhood frames but incomplete with respect to relational frames are found in Shehtman (2005). The proofs of these results are beyond the scope of this book, and so I do not not include them here. The interested reader is invited to consult Litak (2004, 2005) and Shehtman (2005) for further results and an overview of this fascinating literature.

A seemingly simple generalization of neighborhood frames allows us to bypass the incompleteness results mentioned above and to prove a general completeness theorem for *all* classical modal logics. There is an analogous result for relational frames, and a rich mathematical theory of so-called *general* relational frames. See Blackburn et al. (2001, Chap. 5) for an overview of general relational frames and their use in the model theory of normal modal logic. It is beyond the scope of this book to go into the details of this theory. Instead, I show how to adapt the definition of a general relational frame to the neighborhood setting and prove that all classical modal logics are complete with respect to their general neighborhood frame.

Definition 2.43 (*General Neighborhood Frame*) A **general neighborhood frame** is a tuple $\mathcal{F}^g = \langle W, N, \mathcal{A} \rangle$, where W is a non-empty set of states, N is a neighborhood function, and $\mathcal{A}$ is a collection of subsets of W closed under intersections, complements, and the m_N operator.

A valuation $V : \mathsf{At} \to \wp(W)$ is **admissible** for a general frame $\langle W, N, \mathcal{A} \rangle$ provided, for each $p \in \mathsf{At}$, $V(p) \in \mathcal{A}$.

Definition 2.44 (*General Neighborhood Model*) Let $\mathcal{F}^g = \langle W, N, \mathcal{A}\rangle$ be a general neighborhood frame. A general neighborhood model based on $\mathcal{F}^g$ is a tuple $\mathcal{M}^g = \langle W, N, \mathcal{A}, V\rangle$, where V is an admissible valuation.

On general models, truth for the basic modal language is defined as in Definition 1.12. The first observation is that on a general model, the truth set of all formulas is contained in the distinguished collection of propositions.

Lemma 2.16 *Let $\mathcal{M}^g = \langle W, N, \mathcal{A}, V\rangle$ be a general neighborhood model. Then, for each $\varphi \in \mathcal{L}$, $[\![\varphi]\!]_{\mathcal{M}^g} \in \mathcal{A}$.*

Proof The proof is an easy induction over the structure of φ. □

Suppose that **L** is a modal logic containing **E**. It is easy to show that the set $\mathcal{A}_{\mathbf{L}} = \{|\varphi|_{\mathbf{L}} \mid \varphi \in \mathcal{L}\}$ is a Boolean algebra (i.e., closed under intersections and complements) and closed under the m_N operator. A general frame $\mathcal{F}^g$ is called an **L-general frame**, if **L** is valid on $\mathcal{F}^g$. I will show that for any modal logic **L** containing **E**, the canonical general frame is an **L**-general frame. A **general frame** $\langle W_{\mathbf{L}}, N_{\mathbf{L}}, \mathcal{A}_{\mathbf{L}}\rangle$ is said to be the **L-canonical general frame** provided that $\langle W_{\mathbf{L}}, N_{\mathbf{L}},\rangle$ is the minimal **L**-canonical frame (i.e., $W_{\mathbf{L}} = M_{\mathbf{L}}$ and $N_{\mathbf{L}} = N_{\mathbf{L}}^{min}$) and $\mathcal{A}_{\mathbf{L}} = \{|\varphi|_{\mathbf{L}} \mid \varphi \in \mathcal{L}\}$. If **L** is a modal logic containing **E**, then $\mathcal{F}_{\mathbf{L}}^g = \langle W_{\mathbf{L}}, N_{\mathbf{L}}, \mathcal{A}_{\mathbf{L}}\rangle$ is the **L**-canonical frame and $\mathcal{M}_{\mathbf{L}}^g = \langle W_{\mathbf{L}}, N_{\mathbf{L}}, \mathcal{A}_{\mathbf{L}}, V_{\mathbf{L}}\rangle$ is the **L**-canonical general model. Note that the minimal canonical neighborhood function, $N_{\mathbf{L}}^{min}$, is used for *all* classical modal logics. (Compare this to the completeness proofs from Sect. 2.3.2.)

Theorem 2.45 *Suppose that **L** is any logic containing **E**. Then,*

$$\mathcal{F}_{\mathbf{L}}^g \models \mathbf{L}.$$

Proof Suppose that $\mathcal{F}_{\mathbf{L}}^g = \langle W_{\mathbf{L}}, N_{\mathbf{L}}, \mathcal{A}_{\mathbf{L}}\rangle$ is the **L**-canonical general frame. It is a simple exercise to adapt the proof of the Truth Lemma (Lemma 2.13) to prove a Truth Lemma for $\mathcal{M}_{\mathbf{L}}^g$: For all $\varphi \in \mathcal{L}$, $[\![\varphi]\!]_{\mathcal{M}_{\mathbf{L}}^g} = |\varphi|_{\mathbf{L}}$.

Suppose that $\varphi \in \mathbf{L}$, and V is an admissible valuation for $\mathcal{F}_{\mathbf{L}}^g$. We must show that $\mathcal{M}^g = \langle M_{\mathbf{L}}, N_{\mathbf{L}}, \mathcal{A}_{\mathbf{L}}, V\rangle$ validates φ. Since V is admissible for $\mathcal{F}_{\mathbf{L}}^g$, for each propositional letter p_i occurring in φ, $V(p_i) \in \mathcal{A}_{\mathbf{L}}$. Hence, for each p_i (there are only finitely many), $V(p_i) = |\psi_i|_{\mathbf{L}}$ for some formula $\psi_i \in \mathbf{L}$. Let φ' be φ where each p_i is replaced with ψ_i. We prove by induction on φ that

$$[\![\varphi]\!]_{\mathcal{M}^g} = [\![\varphi']\!]_{\mathcal{M}_{\mathbf{L}}^g}.$$

The base case is $\varphi = p$. Then, $\varphi' = \psi$ for some $\psi \in \mathbf{L}$, where $V(p) = |\psi|_{\mathbf{L}} \in \mathcal{A}_{\mathbf{L}}$. Hence,

$$\begin{aligned}\Gamma \in [\![p]\!]_{\mathcal{M}^g} &\text{ iff } \Gamma \in V(p) \in \mathcal{A}_{\mathbf{L}} &&\text{(definition of truth)}\\ &\text{ iff } \Gamma \in |\psi|_{\mathbf{L}} \text{ for some } \psi \in \mathbf{L} &&\text{(since } V(p) = |\psi|_{\mathbf{L}})\\ &\text{ iff } \Gamma \in [\![\psi]\!]_{\mathcal{M}_{\mathbf{L}}^g} &&\text{(Truth Lemma for } \mathcal{M}_{\mathbf{L}}^g)\end{aligned}$$

2.3 The Landscape of Non-normal Modal Logics

The argument for the Boolean cases is as usual. Suppose that φ is of the form $\Box \gamma$. The induction hypothesis is $[\![\gamma]\!]_{\mathcal{M}^g} = [\![\gamma']\!]_{\mathcal{M}_\mathbf{L}^g}$. Then,

$$\begin{aligned} \Gamma \in [\![\Box \gamma]\!]_{\mathcal{M}^g} &\text{ iff } [\![\gamma]\!]_{\mathcal{M}^g} \in N_\mathbf{L}(\Gamma) \quad \text{(Definition of truth)} \\ &\text{ iff } [\![\gamma']\!]_{\mathcal{M}_\mathbf{L}^g} \in N_\mathbf{L}(\Gamma) \quad \text{(Induction hypothesis)} \\ &\text{ iff } \Gamma \in [\![\Box \gamma']\!]_{\mathcal{M}_\mathbf{L}^g} \quad \text{(Definition of truth)} \end{aligned}$$

Suppose that $\varphi \in \mathbf{L}$ and $\mathcal{M}^g = \langle W_\mathbf{L}, N_\mathbf{L}, \mathcal{A}_\mathbf{L}, V \rangle$ is any general model based on the general canonical frame for $\mathbf{L}$. Since $\mathbf{L}$ is closed under uniform substitution, $\varphi' \in \mathbf{L}$ where φ' is φ where each atomic proposition p in φ is replaced with the formula $\psi \in \mathbf{L}$ such that $V(p) = |\psi|_\mathbf{L} \in \mathcal{A}_\mathbf{L}$. Then, by the Truth Lemma for $\mathcal{M}_\mathbf{L}^g$, φ' is valid on $\mathcal{M}_\mathbf{L}^g$. Since $[\![\varphi]\!]_{\mathcal{M}^g} = [\![\varphi']\!]_{\mathcal{M}_\mathbf{L}^g}$, φ is valid on $\mathcal{M}^g$. Hence, $\mathcal{F}_\mathbf{L}^g \models \mathbf{L}$. □

Corollary 2.3 *Every modal logic* $\mathbf{L}$ *extending* $\mathbf{E}$ *is sound and strongly complete with respect to its class of general neighborhood frames.*

Consult Došen (1989) for a more extensive discussion of general frames for neighborhood structures (cf., also, Hansen 2003).

2.4 Computational Issues

Given any logical system (such as neighborhood semantics for modal logic), there are three natural computational problems that arise:

- **Model Checking Problem**: Given a (finitely represented) pointed neighborhood model $\mathcal{M}, w$ and a formula $\varphi \in \mathcal{L}$, does $\mathcal{M}, w \models \varphi$?
- **Satisfiability Problem**: Given a formula φ, is there a model (from some class of models) that satisfies φ? Equivalently, given a formula φ, is φ valid?
- **Model Equivalence Problem**: Given two (finitely represented) pointed neighborhood models $\mathcal{M}, w$ and $\mathcal{N}, v$, do $\mathcal{M}, w$ and $\mathcal{N}, v$ satisfy the same modal formulas?

A variety of algorithms have been proposed (and implemented) to solve the above problems for various classes of relational structures and modal languages. Many of the same ideas can be adapted to the neighborhood setting. In Sect. 2.4.1, I show that the satisfiability problem for non-normal modal logics is decidable, and I discuss the complexity of this problem in Sect. 2.4.2. See, for example, Pauly (2002) for results about the model-checking problem for coalitional logic (cf. Sect. 1.4.5).

2.4.1 Filtrations

Suppose that $\mathcal{M} = \langle W, N, V \rangle$ is a neighborhood model and Σ is a set of formulas from $\mathcal{L}$. For each $w, v \in W$, write $w \sim_\Sigma v$ iff for each $\varphi \in \Sigma$, $w \models \varphi$ iff $v \models \varphi$. In other words, $w \sim_\Sigma v$ iff w and v agree on all formulas in Σ. It is easy to see that $\sim_\Sigma$ is an equivalence relation. For each $w \in W$, let $[w]_\Sigma = \{v \mid w \sim_\Sigma v\}$ be the equivalence class of $\sim_\Sigma$. If $X \subseteq W$, let $[X]_\Sigma = \{[w] \mid w \in X\}$. If Σ is clear from the context, I may leave out the subscripts. A set of sentences Σ is **closed under subformulas** provided that for all $\varphi \in \Sigma$, all subformulas of φ are in Σ. With this notation in place, I can define a filtration.

Definition 2.46 (*Filtration*) Suppose that $\mathcal{M} = \langle W, N, V \rangle$ is a neighborhood model and Σ is a set of sentences closed under subformulas. A **filtration** of $\mathcal{M}$ through Σ is a model $\mathcal{M}^f = \langle W^f, N^f, V^f \rangle$, where

1. $W^f = [W]$
2. For each $w \in W$, for each $\Box\varphi \in \Sigma$, $[\![\varphi]\!]_\mathcal{M} \in N(w)$ iff $[[\![\varphi]\!]_\mathcal{M}] \in N^f([w])$
3. For each $p \in \mathsf{At}$, $V(p) = [V(p)]$

If Σ is finite, then it is easy to see that $\mathcal{M}^f$ will be a finite model. We need only show that this model agrees with $\mathcal{M}$ on all formulas in Σ.

Theorem 2.47 *Suppose that $\mathcal{M}^f = \langle W^f, N^f, V^f \rangle$ is a filtration of $\mathcal{M} = \langle W, N, V \rangle$ through (a subformula closed) set of sentences Σ. Then, for each $\varphi \in \Sigma$,*

$$\mathcal{M}, w \models \varphi \text{ iff } \mathcal{M}^f, [w] \models \varphi.$$

Proof The proof is by induction on the structure of φ. The base case and Boolean connectives are straightforward. (Note that the fact that Σ is closed under subformulas is needed for to apply the induction hypothesis.) I consider only the case for the modal operator. Suppose that φ is of the form $\Box\psi$. Then, since Σ is closed under subformulas and $\Box\psi \in \Sigma$, we have $\psi \in \Sigma$. Then, the induction hypothesis implies that $[[\![\psi]\!]_\mathcal{M}] = [\![\psi]\!]_{\mathcal{M}^f}$. Thus,

$\mathcal{M}, w \models \Box\psi$ iff $[\![\psi]\!]_\mathcal{M} \in N(w)$ (Definition of truth)
  iff $[[\![\psi]\!]_\mathcal{M}] \in N^f([w])$ (Item 2(a) in Definition 2.46)
  iff $[\![\psi]\!]_{\mathcal{M}^f} \in N^f([w])$ (induction hypothesis)
  iff $\mathcal{M}^f, [w] \models \Box\psi$ (Definition of truth)

The argument for the $\Diamond$ operator is similar and left as an exercise for the reader. $\Box$

The **finest filtration** of $\mathcal{M} = \langle W, N, V \rangle$ is defined as follows: $\mathcal{M}^f_{min} = \langle W^f, N^f_{min}, V^f \rangle$, where for all $[w] \in W^f$,

$$N^f_{min}([w]) = \{[[\![\varphi]\!]_\mathcal{M}] \mid [\![\varphi]\!]_\mathcal{M} \in N(w) \text{ and } \Box\varphi \in \Sigma\}.$$

Obviously, the finest filtration is, in fact, a filtration. In general, the finest filtration may not preserve the algebraic properties of the neighborhood functions in a model. For example, the finest filtration may not be closed under intersections or supersets. This can be easily rectified.

2.4 Computational Issues

Lemma 2.17 *Suppose that $\mathcal{M} = \langle W, N, V \rangle$ is closed under supersets and Σ is a subformula closed set of formulas. If $\mathcal{M}^f$ is the finest filtration of $\mathcal{M}$, then $(\mathcal{M}^f)^{mon}$ is a filtration.*

Proof Suppose that $\Box\varphi \in \Sigma$. We prove that $[\![\varphi]\!]_\mathcal{M} \in N(w)$ iff $[[\![\varphi]\!]_\mathcal{M}] \in (N^f_{min}([w]))^{mon}$. Suppose that $[\![\varphi]\!]_\mathcal{M} \in N(w)$. Then, by the definition of $\mathcal{M}^f$, $[[\![\varphi]\!]_\mathcal{M}] \in N^f_{min}([w])$, and so, $[[\![\varphi]\!]_\mathcal{M}] \in (N^f_{min}([w]))^{mon}$.

Now, suppose that $[[\![\varphi]\!]_\mathcal{M}] \in (N^f_{min}([w]))^{mon}$. Then, there is a ψ such that $[[\![\psi]\!]_\mathcal{M}] \in N^f_{min}([w])$ and $[[\![\psi]\!]_\mathcal{M}] \subseteq [[\![\varphi]\!]_\mathcal{M}]$. Since $[[\![\psi]\!]_\mathcal{M}] \in N^f_{min}([w])$, we have $\Box\psi \in \Sigma$ and $[\![\psi]\!]_\mathcal{M} \in N(w)$. We now show that $[\![\psi]\!]_\mathcal{M} \subseteq [\![\varphi]\!]_\mathcal{M}$. Suppose that $v \in [\![\psi]\!]_\mathcal{M}$. Then, since $\psi \in \Sigma$, $[v] \in [[\![\psi]\!]_\mathcal{M}]$. Since $[[\![\psi]\!]_\mathcal{M}] \subseteq [[\![\varphi]\!]_\mathcal{M}]$, we have $[v] \in [[\![\varphi]\!]_\mathcal{M}]$. Theorem 2.47 implies that $v \in [\![\varphi]\!]_\mathcal{M}$. Therefore, since $N(w)$ is closed under supersets and $[\![\psi]\!]_\mathcal{M} \subseteq [\![\varphi]\!]_\mathcal{M}$, we have that $[\![\varphi]\!]_\mathcal{M} \in N(w)$. This completes the proof that $(\mathcal{M}^f)^{mon}$ is a filtration. $\Box$

Exercise 2.23 Prove analogous results for neighborhood structures that (1) are closed under intersections; (2) contain the unit; and (3) are consistent filters.

Theorem 2.47 shows that for any formula $\varphi \in \mathcal{L}$, if it is satisfiable on a neighborhood model, then it is satisfiable on a finite neighborhood model. A careful inspection of the proof of Theorem 2.47 reveals that there is a bound on the size of the finite satisfying model. In fact, this bound is a function of the number of symbols in φ. This means that the satisfiability problem for **E** is decidable (one needs only search a finite number of finite models to determine whether a formula φ has a satisfying model). Lemma 2.17 is needed to show that **EM** is decidable. In fact, the filtration method (using Lemma 2.17 and Exercise 2.23) can be used to prove the following theorem.

Theorem 2.48 *The satisfiability problems for* **E**, **EM**, **EC**, **EMC**, **EN**, **EMN**, **ECN**, *and* **ECMN** *are all decidable.*[12]

2.4.1.1 Logics with Non-iterative Axioms

A formula of modal logic is said to be **non-iterative** provided if it does not contain any modal operators inside the scope of a modal operator. For instance, $\Box p \to q$, $((p \wedge q) \to r) \to ((p \wedge q)\Box \to r)$, and $\Box(p \wedge q) \to (\Box p \wedge \Box q)$ are all examples of non-iterative formulas. Examples of iterative formulas include $\Box p \to \Box\Box p$, $\Diamond p \to \Box\Diamond p$, and $\Box(\Box p \to p)$. Suppose that **L** is a modal logic extending **E** that can be axiomatized by non-iterative axioms. Many of the logics studied in this book are examples of non-iterative logics. For instance, the non-normal logics **E**, **EM**, **EC**, and **EMC** are all non-iterative. Lewis proved that every finitely axiomatizable non-iterative logic is decidable (Lewis 1974, Theorem 2). In this section, I show how

---

[12]See Chellas (1980, Sects. 7.5 and 9.5) for an extended discussion.

to adapt the filtration method from the previous section to prove Lewis's general decidability result.

Suppose that φ is a formula and **L** is a finite axiomatizable non-iterative modal logic extending **E**. Let Σ_φ be the set of all subformulas of φ. A φ-**description** is a set D defined as follows:

$$D = X \cup \{\neg\sigma \mid \sigma \in \Sigma_\varphi - X\}$$

where $X \subseteq \Sigma_\varphi$. So, a φ-description is a set of zero or more subformulas of φ together with negations of all other subformulas of φ. Since Σ_φ is finite, each φ-description D is finite. Furthermore, there are only finitely many φ-descriptions.

For each φ-description D, choose a maximally consistent set Γ_D containing D. Note that by Lindenbaum's Lemma, every **L**-consistent φ-description D is contained in a maximally consistent set (in general, D is contained in many maximally consistent sets). Let $W_\varphi = \{\Gamma_D \mid D \text{ is a } \varphi\text{- description}\}$. We will build a canonical model with W_φ as the set of states.

Recall that if $\vdash_\mathbf{L} \varphi$, then $\varphi \in \Gamma$ for all maximally **L**-consistent sets. Furthermore, the converse is true (Corollary 2.2): If $\varphi \in \Gamma$ for *all* maximally consistent sets Γ, then $\vdash_\mathbf{L} \varphi$. An analogue of Corollary 2.2 holds with respect to W_φ if we restrict attention to Boolean combinations of formulas from Σ_φ (i.e., substitution instances of propositional formulas in which the atomic propositions are replaced by formulas from Σ_φ).

Lemma 2.18 *Suppose that ψ is a truth-functional combination of formulas from Σ_φ. Then, $\vdash_\mathbf{L} \psi$ iff $\psi \in \Gamma_D$ for all $\Gamma_D \in W_\varphi$.*

Proof Suppose that ψ is a truth-functional combination of formulas from Σ_φ. Then, if $\vdash_\mathbf{L} \psi$, then $\psi \in \Gamma$ for all maximally **L**-consistent sets. Thus, in particular, $\psi \in \Gamma_D$ for all $\Gamma_D \in W_\varphi$. To prove the converse, suppose that $\psi \in \Gamma_D$ for all $\Gamma_D \in W_\varphi$. Note that for each φ-description D, since ψ is a Boolean combination of formulas from Σ_φ, it cannot be the case that both $D \cup \{\psi\}$ and $D \cup \{\neg\psi\}$ are **L**-consistent. There are two cases:

Case 1: There is a φ-description D such that $D \cup \{\neg\psi\}$ is **L**-consistent. Then, $D \cup \{\psi\}$ is not **L**-consistent. Thus, $\psi \notin \Gamma_D$. This contradicts the assumption that $\psi \in \Gamma_D$ for all $\Gamma_D \in W_\varphi$.

Case 2: There is no φ-description D such that $D \cup \{\neg\psi\}$ is **L**-consistent. Since ψ is a Boolean combination of formulas from Σ_φ and the φ-descriptions range over all possible truth-value assignments to formulas in Σ_φ, it must be the case that ψ is an instance of a propositional tautology. Hence, $\vdash_\mathbf{L} \psi$. □

Before discussing the main result of this section, I adapt the proof-set notation from the previous section. For any formula ψ, let

$$|\psi|_\varphi = \{\Gamma_D \in W_\varphi \mid \psi \in \Gamma_D\}.$$

2.4 Computational Issues

A key observation is that every element of W_φ can be associated with a modal formula. Fix an enumeration of the formulas of the language of the modal logic **L**. For each $\Gamma_D \in W_\varphi$, let λ_D be the first (in the enumeration of all formulas) conjunction of all the formulas in D. The formula λ_D is called the **label** of Γ_D. Using these labels, we can associate a formula with every subset of W_φ. If $X \subseteq W_\varphi$ and $X \neq \emptyset$, then let $\lambda_X = \bigvee_{\Gamma_D} \lambda_D$. Let $\lambda_\emptyset = \varphi \wedge \neg\varphi$. Two immediate consequences of these definitions are:

1. For each $X \subseteq W_\varphi$, λ_X is a Boolean combination of formulas from Σ_φ; and
2. For each $X \subseteq W_\varphi$, $|\lambda_X|_\varphi = X$.

We can now define a finite φ-canonical model.

Definition 2.49 (φ-*Canonical Model*) Suppose that φ is a modal formula. A φ-**Canonical Model** is a structure $\mathcal{M}_\varphi = \langle W_\varphi, N_\varphi, V_\varphi \rangle$, where W_φ is defined as above. The neighborhood function $N_\varphi : W_\varphi \to \wp(\wp(W_\varphi))$ is defined as follows: For all $\Gamma_D \in W_\varphi$,
$$N(\Gamma_D) = \{X \mid X \subseteq W_\varphi \text{ and } \Box\lambda_X \in \Gamma_D\}$$

The valuation $V_\varphi : \mathsf{At} \to \wp(W_\varphi)$ is defined as $V_\varphi(p) = |p|_\varphi$ for all $p \in \mathsf{At}$.

Note that $\mathcal{M}_\varphi$ is indeed a filtration (Definition 2.46) of the canonical model for **L**. However, there is a stronger version of Theorem 2.47.

Proposition 2.7 *Suppose that φ is a modal formula and that $\mathcal{M}_\varphi = \langle W_\varphi, N_\varphi, V_\varphi \rangle$ is a φ-canonical model for a consistent modal logic* **L** *containing* **E**. *Then,*

1. *If ψ is a Boolean combination of formulas $\psi_1, \ldots, \psi_k$ such that $[\![\psi_i]\!]_{\mathcal{M}_\varphi} = |\psi_i|_\varphi$ for $i = 1, \ldots, k$, then $[\![\psi]\!]_{\mathcal{M}_\varphi} = |\psi|_\varphi$; and*
2. *If (a) ψ is a Boolean combination of formulas from Σ_φ and that (b) $[\![\psi]\!]_{\mathcal{M}_\varphi} = |\psi|_\varphi$, then $[\![\Box\psi]\!]_{\mathcal{M}_\varphi} = |\Box\psi|_\varphi$.*

Proof The proof of item 1 is straightforward, and so, it is left as an exercise for the reader. We prove item 2. Suppose that (a) ψ is a Boolean combination of formulas from Σ_φ and that (b) $[\![\psi]\!]_{\mathcal{M}_\varphi} = |\psi|_\varphi$. Then,

$$\begin{aligned}[][\![\Box\psi]\!]_{\mathcal{M}_\varphi} &= \{\Gamma_D \mid [\![\psi]\!]_{\mathcal{M}_\varphi} \in N_\varphi(\Gamma_D)\} \\ &= \{\Gamma_D \mid \Box\lambda_{[\![\psi]\!]_{\mathcal{M}_\varphi}} \in \Gamma_D\} \\ &= |\Box\lambda_{[\![\psi]\!]_{\mathcal{M}_\varphi}}|_\varphi \end{aligned}$$

To complete the proof, we must show that $|\Box\lambda_{[\![\psi]\!]_{\mathcal{M}_\varphi}}|_\varphi = |\Box\psi|_\varphi$. Now by assumption (b), $[\![\psi]\!]_{\mathcal{M}_\varphi} = |\psi|_\varphi$ and property 2 of labels, we have that $|\lambda_{[\![\psi]\!]_{\mathcal{M}_\varphi}}| = [\![\psi]\!]_{\mathcal{M}_\varphi} = |\psi|_\varphi$. Thus, $|\lambda_{[\![\psi]\!]_{\mathcal{M}_\varphi}}| = |\psi|_\varphi$. This means that $\lambda_{[\![\psi]\!]_{\mathcal{M}_\varphi}} \leftrightarrow \psi \in \Gamma_D$ for all $\Gamma_D \in W_\varphi$. By assumption (a) and the definition of a label, the formula $\lambda_{[\![\psi]\!]_{\mathcal{M}_\varphi}} \leftrightarrow \psi$ is a Boolean combination of formulas from Σ_φ. By Lemma 2.18, $\vdash_\mathbf{L} \lambda_{[\![\psi]\!]_{\mathcal{M}_\varphi}} \leftrightarrow \psi$. Since **L** contains the rule (**RE**), we have that $\vdash_L \Box \lambda_{[\![\psi]\!]_{\mathcal{M}_\varphi}} \leftrightarrow \Box \psi$. Applying Lemma 2.18 again, we have that $\Box \lambda_{[\![\psi]\!]_{\mathcal{M}_\varphi}} \leftrightarrow \Box \psi \in \Gamma_D$ for all $\Gamma_D \in W_\varphi$. This implies that $|\Box \lambda_{[\![\psi]\!]_{\mathcal{M}_\varphi}}|_\varphi = |\Box \psi|_\varphi$, as desired. □

Propositions 2.18 and 2.7 lead to the following key observation.

Proposition 2.8 *1. Suppose that **L** is a consistent modal logic extending **E** with non-iterative axioms. If φ is any formula and $\vdash_L \psi$, then ψ is valid on the φ-canonical frame $\langle W_\varphi, N_\varphi \rangle$.*
2. If $\nvdash_L \varphi$, then φ is not valid on the φ-canonical frame $\langle W_\varphi, N_\varphi \rangle$.

Proof Proof of item 1: Suppose that ψ is an instance of a non-iterative axiom of **L**. We must show that ψ is valid on the φ-canonical frame $\langle W_\varphi, N_\varphi \rangle$. Let $\mathcal{M}' = \langle W_\varphi, N_\varphi, V \rangle$ be any model based on $\langle W_\varphi, N_\varphi \rangle$. We must show that $[\![\psi]\!]_{\mathcal{M}'} = W_\varphi$. Let $p_1, \ldots, p_k$ be the atomic propositions occurring in ψ, and let σ be any substitution in which for all $i = 1, \ldots, k$, $\sigma(p_i) = \lambda_{[\![p_i]\!]_{\mathcal{M}'}}$. Then, ψ^σ is the formula ψ in which each atomic proposition p_i is replaced by the label of $[\![p_i]\!]_{\mathcal{M}'}$. Since ψ is an instance of an non-iterative axiom, we have that ψ^σ is also a substitution instance of a non-iterative axiom. Hence, $\vdash_L \psi^\sigma$. By Lemma 2.11, $\psi^\sigma \in \Gamma$ for all maximally **L**-consistent sets. Thus, $\psi^\sigma \in \Gamma_D$ for all $\Gamma_D \in W_\varphi$. This means that $|\psi^\sigma|_\varphi = W_\varphi$. Furthermore, since ψ is a non-iterative formula and for each $i = 1, \ldots, k$, $[\![\lambda_{[\![p_i]\!]_{\mathcal{M}'}}]\!]_{\mathcal{M}_\varphi} = |\lambda_{[\![p_i]\!]_{\mathcal{M}'}}|_\varphi$, Proposition 2.7 implies that $[\![\psi^\sigma]\!]_{\mathcal{M}_\varphi} = |\psi^\sigma|_\varphi$. Therefore, $[\![\psi^\sigma]\!]_{\mathcal{M}_\varphi} = W_\varphi$. A simple induction on formulas shows that $[\![\psi^\sigma]\!]_{\mathcal{M}_\varphi} = [\![\psi]\!]_{\mathcal{M}'}$ (see Exercise 1.5). Hence, $[\![\psi]\!]_{\mathcal{M}'} = W_\varphi$, as desired.

Proof of item 2: Suppose that $\nvdash_L \varphi$. Then, since φ is trivially a Boolean combination of formulas from Σ_φ, by Lemma 2.18, we have that $[\![\varphi]\!]_{\mathcal{M}_\varphi} \neq W_\varphi$. Thus, φ is not valid on $\langle W_\varphi, N_\varphi \rangle$. □

This proposition shows that any non-iterative modal logic containing **E** is weakly complete (Remark 2.35). Furthermore, since there is a bound on the size of W_φ, which can be calculated from φ, we only need to check finitely many[13] models to determine if φ is derivable in **L**.

Theorem 2.50 (Lewis 1974) *Every non-iterative modal logic **L** containing **E** is weakly complete and decidable.*

There are two ways to extend this result. First, Lewis (1974) worked with a more general modal language including modalities of arbitrary arity. It is not hard to adapt the argument of this section to this more general setting. Lewis's theorem, then, gives a weak completeness and decidability result for the minimal conditional logic of sphere model. That is, the following axiom system (expressed using the comparative possibility modality $\preceq$ discussed in Sect. 1.4.3) is sound and complete for the class of all sphere frames.

---

[13] Of course, there are infinitely many variations of the finite φ-canonical model. However, we can ignore irrelevant differences, such as isomorphic copies or models that differ in their interpretation of formulas not among the subformulas of φ.

2.4 Computational Issues 75

(taut) All instances of propositional tautologies
(trans) $((\varphi \preceq \psi) \wedge (\psi \preceq \chi)) \rightarrow (\varphi \preceq \chi)$
(cons) $(\varphi \preceq \psi) \vee (\psi \preceq \varphi)$
(dis) $((\varphi \vee \psi) \preceq \chi) \rightarrow ((\varphi \preceq \chi) \vee (\psi \preceq \chi))$
(RE) $\dfrac{\varphi \leftrightarrow \psi}{\chi \leftrightarrow \chi'}$ where χ' is χ with one or more occurrences of φ replaced with ψ.

A second generalization comes from Surendonk (2001) who showed that the argument in this section can be adapted to prove strong completeness for non-iterative logics. See, also, Schröder (2006), Pattinson and Schröder (2006) and Schröder and Pattinson (2007) for further results about non-iterative modal logics.

2.4.2 Complexity

This brief section assumes that the reader has some familiarity with computation complexity theory. As a reminder, a problem is in **NP** provided that it can be solved by a non-deterministic Turing machine that is guaranteed to halt after a number of steps that is polynomial in the size of the input. A problem is **NP**-complete provided that it is in **NP** and that any other problem in **NP** can be polynomially reduced to it. Intuitively, a problem is **NP**-complete if it is the "hardest" problem solved by a non-deterministic Turing machine in a polynomial amount of time. The other complexity class that will be mentioned is **PSPACE**. A problem is in **PSPACE** provided that it can be decided by a deterministic Turing machine that uses at most a polynomial (in the size of the input) amount of space (i.e., tape cells). A problem is **PSPACE**-complete provided that it is in **PSPACE** and that every problem in **PSPACE** can be reduced to it. It is known that every problem in **NP** is also in **PSPACE**, but the converse is still open (i.e., is every problem in **PSPACE** also in **NP**?). However, it is widely believed that **PSPACE**-complete problems are "harder" than **NP**-complete problems.

Using the above two complexity classes, we can classify how hard it is to solve the satisfiability problem for different logical systems. The satisfiability problem for propositional logic is **NP**-complete. On the other hand, as is well known, the satisfiability problem for first-order logic is undecidable (see Enderton 2001, Chap. 3 for a discussion). The satisfiability problem for normal modal logic (i.e., with respect to relational models) is **PSPACE**-complete. Interestingly, the satisfiability problem for some classes of relational models is "easier". For example, the satisfiability problem for the class of all relational frames in which the relations are equivalence relations is **NP**-complete. Marx (2007) offers a more in-depth discussion of complexity issues in modal logic.[14] What about the complexity of the satisfiability problem for neighborhood models?

---

[14]Consult Halpern and Rêgo (2007) and Spaan (1993) for discussions of the curious fact that the satisfiability problem for modal logics seems to be either **NP**-complete or **PSPACE**-hard.

The algorithm outlined in Sect. 2.4.1, which searches all finite models of a bounded size to find a satisfying model runs in double exponential time (this follows from the fact that there are exponentially many subformulas of a given formula). Vardi (1989) showed that there is a much more efficient algorithm, proving that the satisfiability problem with respect to all neighborhood frames is **NP**-complete. As in the case with relational models, this result is sensitive to the properties of the neighborhood function. Vardi (1989) also showed that the satisfiability problem for the class of neighborhood models that are closed under intersections (i.e., the class of models for the logic **EC**) is **PSPACE**-complete. Interestingly, Allen (2005) later showed that there are additional non-normal modal logics with **PSPACE**-complete satisfiability problems. Allen studied logics inspired by the generalized relational models discussed in Sect. 2.2.2. For each $n \geq 2$, let (C^n) be the follow axiom:

$$(C^n) \quad \bigwedge_{i=1}^{n} \Box \varphi_i \rightarrow \Box \bigvee_{1 \leq k, l \leq n,\ k \neq l} (\varphi_k \wedge \varphi_l).$$

Let **EMNC**^{n} be the logic extending **EMN** with instances of (C^n). Allen (2005) proved that for each $n \geq 2$, the satisfiability problem for **EMNC**^{n} is **PSPACE**-complete.

2.4.3 Proof Theory for Non-normal Modal Logics

While there is an extensive body of research focused on developing proof calculi for normal modal logics,[15] there has been much less work developing proof calculi for non-normal modal logics. Notable exceptions include Chap. 6 in Fitting (1983), the labeled tableaux systems for non-normal modal logics in Governatori and Luppi (2000) and a series of recent papers developing sequent calculi for non-normal modal logics (Gilbert and Maffezioli 2015; Girlando et al. 2016).

In this section, I briefly introduce *sequents* for non-normal modal logics. My aim in this section is to use these sequents to further illustrate the issues that arose when discussing decidability (Sect. 2.4.1) and complexity (Sect. 2.4.2), rather than providing a complete introduction to sequent calculi for non-normal modal logics.

Before defining a sequent, I discuss an elegant characterization of some non-normal modal logics (cf. Definition 2.26) using inference rules. Consider the following inference rules (where k ranges over the non-negative integers):

$$(R_k) \quad \frac{(\varphi_1 \wedge \cdots \wedge \varphi_k) \rightarrow \psi}{(\Box \varphi_1 \wedge \cdots \wedge \Box \varphi_k) \rightarrow \Box \psi}.$$

When $k = 0$ (so there are no antecedents), the above rule reduces to the rule of Necessitation (Nec). When $k = 1$, the above rule is the monotonicity rule (RM) discussed in Sect. 2.3. Then, Lemma 2.6 can be rephrased as:

---

[15] See Fitting (2006), Wansing (1998), and Negri (2011) for surveys of this literature.

2.4 Computational Issues

A modal logic **S** is monotonic iff (R_1) is a derived rule.
This suggests the following characterization of normal and regular modal logics.

Proposition 2.9 *Suppose that* **S** *is a modal logic that contains* **E**. *Then,*

- **S** *is regular iff* (R_k) *is a derived rule for all* $k \geq 1$.
- **S** *is normal iff* (R_k) *is a derived rule for all* $k \geq 0$.

Exercise 2.24 Prove Proposition 2.9.

A careful examination of the above proof suggests a sequent-based proof calculus for non-normal modal logic. I start by reminding the reader of the definition of a sequent and the sequent rules for propositional logic.

Definition 2.51 (*Sequent*) A **sequent** is a structure $\Gamma \Rightarrow \Delta$, where Γ and Δ are finite sequences[16] of modal formulas. We write Γ, φ to denote the sequence of formulas in which φ is the last element (similarly, φ, Γ is a sequence in which φ is the first element). A sequent is **valid** on a class of neighborhood frames F if $\bigwedge \Gamma \to \bigvee \Delta$ is valid on F.

As the reader is invited to verify, each of the following rules preserves validity of the sequents and, together, form a complete system for all propositional tautologies.

Definition 2.52 (*Propositional Sequent Rules*)

$$\frac{}{\Gamma, p \Rightarrow p, \Delta}(\text{axiom})$$

$$\frac{\Gamma \Rightarrow \Delta}{\Gamma' \Rightarrow \Delta'}(\text{perm}) \quad \frac{\Gamma \Rightarrow \Delta}{\Gamma, \Pi \Rightarrow \Delta, \Lambda}(\text{weak}) \quad \frac{\Gamma \Rightarrow \Delta, \varphi \quad \varphi, \Pi \Rightarrow \Lambda}{\Gamma, \Pi \Rightarrow \Delta, \Lambda}(\text{cut})$$

where Γ' and Δ' are permutations of Γ and Δ, respectively.

$$\frac{\Gamma, \varphi, \psi \Rightarrow \Delta}{\Gamma, \varphi \wedge \psi \Rightarrow \Delta}(\wedge \text{L}) \quad \frac{\Gamma \Rightarrow \varphi, \Delta \quad \Gamma \Rightarrow \psi, \Delta}{\Gamma \Rightarrow \varphi \wedge \psi, \Delta}(\wedge \text{R})$$

$$\frac{\Gamma, \varphi \Rightarrow \Delta \quad \Gamma, \psi \Rightarrow \Delta}{\Gamma, \varphi \vee \psi \Rightarrow \Delta}(\vee \text{L}) \quad \frac{\Gamma \Rightarrow \varphi, \psi, \Delta}{\Gamma \Rightarrow \varphi \vee \psi, \Delta}(\vee \text{R})$$

$$\frac{\Gamma \Rightarrow \varphi, \Delta}{\Gamma, \neg \varphi \Rightarrow \Delta}(\neg \text{L}) \quad \frac{\Gamma \Rightarrow \neg \varphi, \Delta}{\Gamma, \varphi \Rightarrow \Delta}(\neg \text{R})$$

The above rules can be used to reduce sequents containing complex modal formulas to a sequent of the following form:

$$\mathbf{p}, \Box \varphi_1, \ldots, \Box \varphi_k \Rightarrow \mathbf{q}, \Box \psi_1, \ldots, \Box \psi_m,$$

---

[16] I have been using capital Greek letters to denote sets of formulas (c.f. Sect. 2.3.2). For the purposes of this section, it does not matter much whether the components of a sequent are sets or sequences. However, it is standard practice to define sequents using *sequences* of formulas. So, I will adopt the convention that capital Greek letters denote sequences of formulas in this section.

where **p** and **q** are sequences of propositional variables. The main question when developing sequent rules for non-normal modal logics is how to further reduce the above sequent. It is simplest to start with the logic **EM**. The key observation for this logic is from van Benthem (2010):

Proposition 2.10 (van Benthem 2010) *The sequent*

$$\mathbf{p}, \Box\varphi_1, \ldots, \Box\varphi_k \Rightarrow \mathbf{q}, \Box\psi_1, \ldots, \Box\psi_m$$

is valid on the class of monotonic neighborhood frames if, and only if, either

- **p** *and* **q** *overlap, or*
- *there is some* $1 \leq i \leq k$ *and* $1 \leq j \leq m$ *such that* $\varphi_i \Rightarrow \psi_j$ *is valid on the class of monotonic neighborhood frames.*

Proof I show that if there is no overlap between **p** and **q** and $\varphi_i \Rightarrow \psi_j$ is not valid for each $1 \leq i \leq k$ and $1 \leq j \leq m$, then $\mathbf{p}, \Box\varphi_1, \ldots, \Box\varphi_k \Rightarrow \mathbf{q}, \Box\psi_1, \ldots, \Box\psi_m$ is not valid. The remaining cases of the proof are left to the reader. Suppose that for each i, j (to simplify the proof, I will write "for each i, j" instead of "for each $1 \leq i \leq k$ and $1 \leq j \leq m$"), $\varphi_i \Rightarrow \psi_j$ is not valid. Then, for each i, j, there is a neighborhood model $\mathcal{M}_{ij} = \langle W_{ij}, N_{ij}, V_{ij}\rangle$ with a state $w_{ij} \in W_{ij}$ such that $\mathcal{M}_{ij}, w_{ij} \models \varphi_i \wedge \neg\psi_j$. I will show that $\mathbf{p}, \Box\varphi_1, \ldots, \Box\varphi_k \Rightarrow \mathbf{q}, \Box\psi_1, \ldots, \Box\psi_m$ is not valid. Without loss of generality, we can assume that the sets W_{ij} are pairwise disjoint. Let $\mathcal{M} = \langle W, N, V\rangle$ be a neighborhood model, where

- $W = \bigcup_{i,j} W_{ij} \cup \{w^*\}$ (with w^* a new state not in any W_{ij}).
- Define $V : \mathsf{At} \to \wp(W)$ as follows: Let $V_0 : \mathsf{At} \to \wp(\bigcup_{i,j} W_{ij})$ be the function where for all $p \in \mathsf{At}$, $V(p) = \bigcup_{i,j} V_{ij}(p)$. Then, V is the function where for all $p \in \mathsf{At}$:

$$V(p) = \begin{cases} V_0(p) \cup \{w^*\} & \text{if p is on the list } \mathbf{p} \\ V_0(p) & \text{otherwise.} \end{cases}$$

- Define $N : W \to \wp(\wp(W))$ as follows: For all $w \in \bigcup_{i,j} W_{ij}$, $N(w) = \{X \mid X \subseteq W$ and there is a $Y \in N_{ij}(w)$ such that $Y \subseteq X\}$, where i, j are the unique indices such that $w \in W_{ij}$. The neighborhood at w^* is defined as follows:

$$N(w^*) = \{X \mid X \subseteq W \text{ and } w^* \in X\}.$$

It is immediate from the definition that the above model $\mathcal{M} = \langle W, N, V\rangle$ is a monotonic neighborhood model. Furthermore, by construction we have that $\mathcal{M}, w^* \models \bigwedge_{1 \leq i \leq k} \Box\varphi_i$ and $\mathcal{M}, w^* \not\models \bigvee_{1 \leq j \leq m} \Box\psi_j$. It follows from this observation and the definition of the above model that the sequent $\mathbf{p}, \Box\varphi_1, \ldots, \Box\varphi_k \Rightarrow \mathbf{q}, \Box\psi_1, \ldots, \Box\psi_m$ is not valid. □

Proposition 2.10 justifies the following rule for **EM**:

2.4 Computational Issues

$$\frac{\Gamma, \varphi \Rightarrow \Delta, \psi}{\Gamma, \Box\varphi \Rightarrow \Delta, \Box\psi}(\Box M)$$

Sequent rules for other non-normal modal logics can be developed in a similar manner. The minimal non-normal modal logic **E** requires a simple modification of Proposition 2.10.

Proposition 2.11 *The sequent*

$$\mathbf{p}, \Box\varphi_1, \ldots, \Box\varphi_k \Rightarrow \mathbf{q}, \Box\psi_1, \ldots, \Box\psi_m$$

is valid on the class of neighborhood frames if, and only if, either

- **p** *and* **q** *overlap, or*
- *there is some $1 \le i \le k$ and $1 \le j \le m$ such that both $\varphi_i \Rightarrow \psi_j$ and $\psi_j \Rightarrow \varphi_i$ are valid on the class of neighborhood frames.*

Exercise 2.25 Prove Proposition 2.11. (Hint: Adapt the proof of Proposition 2.10.)

The corresponding sequent rule for **E** is:

$$\frac{\Gamma, \varphi \Rightarrow \Delta, \psi \quad \Gamma, \psi \Rightarrow \Delta, \varphi}{\Gamma, \Box\varphi \Rightarrow \Delta, \Box\psi}(\Box E)$$

Finally, there is an analogous result for the class of neighborhood frames closed under intersections.

Proposition 2.12 *The sequent*

$$\mathbf{p}, \Box\varphi_1, \ldots, \Box\varphi_k \Rightarrow \mathbf{q}, \Box\psi_1, \ldots, \Box\psi_m$$

is valid on the class of neighborhood frames that are closed under finite intersections if, and only if, either

- **p** *and* **q** *overlap, or*
- *there is some $1 \le i, j \le k$ such that $\mathbf{p}, \Box\varphi_1, \ldots, \Box\varphi_{k'}, \Box(\varphi_i \wedge \varphi_j) \Rightarrow \mathbf{q}, \Box\psi_1, \ldots \Box\psi_m$ is valid on the class of neighborhood frames closed under finite intersections (where $\Box\varphi_1, \ldots, \Box\varphi_{k'}$ is the sequence $\Box\varphi_1, \ldots, \Box\varphi_k$ without $\Box\varphi_i$ and $\Box\varphi_j$).*

Exercise 2.26 Prove Proposition 2.12.

The resulting sequent rule for the logic **EC** is:

$$\frac{\Gamma, \Box(\varphi \wedge \psi) \Rightarrow \Delta}{\Gamma, \Box\varphi, \Box\psi \Rightarrow \Delta}(\Box C)$$

There is an important difference between the sequent rules for **E** and **EM** and the one for **EC**. Reading the rules from the bottom up, the first two rules reduce the complexity of formulas in a sequent (i.e., the rules remove the '$\Box$' from two formulas in a sequent). The sequent for **EC** behaves differently. This rule does not simplify any formulas, but, rather, reduces the length of the antecedent of the sequent. This difference has ramifications for the complexity of the proof search problem (given a formula φ, determine if there is a proof of φ). Indeed, this difference explains, in part, why the complexity of the satisfiability problem for **E** and **EM** is **NP**-complete, while it is **PSPACE**-complete for **EC**.

This is just the first steps towards a complete sequent system for non-normal modal logics. Readers interested in developing these ideas further are invited to consult Negri (2016), and the references therein.

2.5 Frame Correspondence

A central topic in the model theory of modal logic is *correspondence theory* (cf. Blackburn et al. 2001, Chap. 3). The aim of this theory is to identify (and characterize) modal formulas that correspond to interesting properties of relational frames. For instance, it is well known that a relational frame $\mathcal{F} = \langle W, R \rangle$ validates $\Box \varphi \to \varphi$ iff R is reflexive. The modal formula $\Box \varphi \to \varphi$ is said to **correspond** to the reflexivity property. See Appendix A for an introduction to correspondence theory with respect to relational structures.

Many of the ideas of correspondence theory can be adapted to the neighborhood setting. In this case, modal formulas express properties of neighborhood functions.

Definition 2.53 A modal formula $\varphi \in \mathcal{L}$ **defines a property** P of neighborhood functions if any neighborhood frame $\mathcal{F} = \langle W, N \rangle$ has property P iff $\mathcal{F}$ validates φ.

There are some important differences when formulas are interpreted on neighborhood frames instead of relational frame. The first difference is that formulas corresponding to the same property on relational frames may correspond to different properties on neighborhood frames. For example, consider the formulas $\Diamond \top$ and $\Box \varphi \to \Diamond \varphi$. On relational frames, these formulas both define the same property: seriality (a relation R is serial provided that for each state w, there is a state v such that $w\ R\ v$).[17] However, on the class of neighborhood frames, these formulas express different properties. The first formula $\Diamond \top$ is easily seen to express the fact that the empty set is not an element of the neighborhoods. That is, $\Diamond \top$ is valid on a neighborhood frame $\mathcal{F}$ iff the empty set is not an element of any neighborhood (the proof follows immediately from the definition of truth of modal formulas). The second formula expresses a different property of neighborhood functions.

Lemma 2.19 *Let $\mathcal{F} = \langle W, N \rangle$ be a neighborhood frame. Then, $\mathcal{F} \models \Box \varphi \to \Diamond \varphi$ iff $\mathcal{F}$ is proper (i.e., if $X \in N(w)$, then $X^C \notin N(w)$).*

---

[17]This makes sense since these formulas are semantically equivalent on the class of relational frames (i.e., they are true at exactly the same points in all relational models).

2.5 Frame Correspondence

Proof The right-to-left direction is straightforward. For the left-to-right direction, suppose that $\mathcal{F} = \langle W, N \rangle$ is not proper. Then, there is a state $w \in W$ and a set $X \subseteq W$ such that $X \in N(w)$ and $X^C \in N(w)$. Define a model $\mathcal{M} = \langle W, N, V \rangle$ with $V(p) = X$. Then, by definition, $\mathcal{M}, w \models \Box p$, and since $[\![\neg p]\!]_\mathcal{M} = X^C \in N(w)$, we have $\mathcal{M}, w \models \neg \Diamond p$. Thus, $\mathcal{M}, w \not\models \Box p \to \neg \Box \neg p$; and so, $\Box \varphi \to \Diamond \varphi$ is not valid on $\mathcal{F}$. □

A second difference is that modal formulas that are valid on all relational frames correspond to non-trivial properties of neighborhood functions.

Lemma 2.20 *Suppose that $\mathcal{F} = \langle W, N \rangle$ is a neighborhood frame. Then, $\mathcal{F} \models \Box \varphi \land \Box \psi \to \Box(\varphi \land \psi)$ iff $\mathcal{F}$ is closed under finite intersections.*[18]

Proof Suppose that $\mathcal{F} = \langle W, N \rangle$ is a neighborhood frame that is closed under finite intersections. We must show $\mathcal{F} \models \Box \varphi \land \Box \psi \to \Box(\varphi \land \psi)$. Let $\mathcal{M} = \langle W, N, V \rangle$ be any model based on $\mathcal{F}$ and $w \in W$. Suppose that $\mathcal{M}, w \models \Box \varphi \land \Box \psi$. Then, $[\![\varphi]\!]_\mathcal{M} \in N(w)$ and $[\![\psi]\!]_\mathcal{M} \in N(w)$. Since $N(w)$ is closed under finite intersections, $[\![\varphi]\!]_\mathcal{M} \cap [\![\psi]\!]_\mathcal{M} \in N(w)$. Hence, $[\![\varphi \land \psi]\!]_\mathcal{M} \in N(w)$ and, therefore, $\mathcal{M}, w \models \Box(\varphi \land \psi)$.

Suppose that $\mathcal{F} = \langle W, N \rangle$ is not closed under finite intersections. Then, $\langle W, N \rangle$ is not closed under binary interactions (see Lemma 1.2). Thus, there is a state $w \in W$ and two sets Y and Y' such that $Y, Y' \in N(w)$, but $Y \cap Y' \notin N(w)$. Define a valuation function so that $V(p) = Y$ and $V(q) = Y'$. This implies that $\mathcal{M}, w \models \Box p \land \Box q$. However, since $Y \cap Y' \notin N(w)$, $\mathcal{M}, w \not\models \Box(p \land q)$. □

Lemma 2.21 *Suppose that $\mathcal{F} = \langle W, N \rangle$ is a neighborhood frame. Then, $\mathcal{F} \models \Box(\varphi \land \psi) \to \Box \varphi \land \Box \psi$ iff $\mathcal{F}$ is closed under supersets.*

Proof The right-to-left direction is left as an exercise for the reader. Suppose that $\mathcal{F} = \langle W, N \rangle$ is not closed under supersets. Then, there are sets X and Y such that $X \subseteq Y$, $X \in N(w)$ but $Y \notin N(w)$. Define a valuation V such that $V(p) = X$ and $V(q) = Y$. Then, since $X \subseteq Y$, $[\![p \land q]\!]_\mathcal{M} = X \in N(w)$. Hence, $\mathcal{M}, w \models \Box(p \land q)$. However, since, $[\![q]\!]_\mathcal{M} = Y \notin N(w)$, we have $\mathcal{M}, w \not\models \Box q$. Hence, $\mathcal{M}, w \not\models \Box p \land \Box q$. □

Lemma 2.22 *Suppose that $\mathcal{F} = \langle W, N \rangle$ is a neighborhood frame. Then, $\mathcal{F} \models \Box \top$ iff $\mathcal{F}$ contains the unit.*

Proof Left as an exercise for the reader. □

I conclude this brief introduction to correspondence theory by identifying properties of neighborhood functions that correspond to well-studied modal formulas.

Lemma 2.23 *Suppose that $\mathcal{F} = \langle W, N \rangle$ is a neighborhood frame such that for each $w \in W$, $N(w) \neq \emptyset$.*

---

[18] Recall that a frame $\mathcal{F} = \langle W, N \rangle$ is said to be closed under finite intersection provided that for all $w \in W$, $N(w)$ is closed under finite intersections. See the discussion after Remark 1.10.

1. $\mathcal{F} \models \Box\varphi \to \varphi$ iff for each $w \in W$, $w \in \cap N(w)$.
2. $\mathcal{F} \models \Box\varphi \to \Box\Box\varphi$ iff for each $w \in W$, if $X \in N(w)$, then $m_N(X) \in N(w)$ (recall that $m_N(X) = \{v \mid X \in N(v)\}$).

Proof Suppose that $\mathcal{F} = \langle W, N \rangle$ is a neighborhood frame. Suppose that for each $w \in W$, $w \in \cap N(w)$. Let $\mathcal{M}$ be any model based on $\mathcal{F}$ and $w \in W$. Suppose that $\mathcal{M}, w \models \Box\varphi$. Then, $[\![\varphi]\!]_\mathcal{M} \in N(w)$. Since $w \in \cap N(w) \subseteq [\![\varphi]\!]_\mathcal{M}$, $w \in [\![\varphi]\!]_\mathcal{M}$. Hence, $\mathcal{M}, w \models \varphi$. As for the converse, suppose that $w \notin \cap N(w)$. Since $N(w) \neq \emptyset$, there is an $X \in N(w)$ (note that X may be empty) such that $w \notin X$; otherwise, $w \in \cap N(w)$. Define a valuation V such that $V(p) = X$. Then, it is easy to see that $\mathcal{M}, w \models \Box p$, but $\mathcal{M}, w \not\models p$.

Suppose that for each $w \in W$, if $X \in N(w)$, then $\{v \mid X \in N(v)\} \in N(w)$. Suppose that $\mathcal{M}$ is any model based on $\mathcal{F}$ and $\mathcal{M}, w \models \Box\varphi$. Then, $[\![\varphi]\!]_\mathcal{M} \in N(w)$. Therefore, by assumption $\{v \mid [\![\varphi]\!]_\mathcal{M} \in N(v)\} \in N(w)$. Since $[\![\Box\varphi]\!]_\mathcal{M} = \{v \mid [\![\varphi]\!]_\mathcal{M} \in N(w)\}$, $\mathcal{M}, w \models \Box\Box\varphi$. For the other direction, suppose that there is some state $w \in W$ and set X such that $X \in N(w)$, but $\{v \mid X \in N(v)\} \notin N(w)$. Then, define a valuation V such that $V(p) = X$. It is easy to verify that $\mathcal{M}, w \models \Box p$, but $\mathcal{M}, w \not\models \Box\Box p$. □

Exercise 2.27 Find properties on frames that are defined by the following formulas:

1. $\neg\Box\varphi \to \Box\neg\Box\varphi$
2. $\Box\varphi \vee \Box\neg\varphi$
3. $\Diamond\Box\varphi \to \Box\Diamond\varphi$

There is much more to say about correspondence theory with respect to neighborhood frames. In particular, the Sahlqvist Theorem (Sahlqvist 1975) provides a syntactic definition of a class of modal formulas, each of which corresponds to a first-order property of a relational frame.[19] An interesting line of research is to explore generalizations of the Sahlqvist theorem and related algorithms (Conradie et al. 2006) for finding first-order correspondents to the class of neighborhood frames. One approach is to first translate neighborhood models and the basic modal language into a special class of relational models with an appropriate modal language (see Sect. 2.6.2 for details of such a translation). The Sahlqvist Theorem can then be applied to this special class of relational models and modal language. Consult Kracht and Wolter (1999) and Hansen (2003) for details of this approach. A second approach is to explore different generalizations of the Sahlqvist Theorem. Consult Palmigiano et al. (2016) for a generalization of the Sahlqvist Theorem to regular modal logics (Definition 2.26).

2.6 Translations

Neighborhood models generalize relational models by replacing a relation between worlds ($R \subseteq W \times W$) with a relation between worlds and sets of worlds ($N \subseteq$

---

[19] See Sects. 3.5–3.7 in Blackburn et al. (2001) for details.

2.6 Translations

$R \times \wp(W)$). In Sect. 2.2.1, we saw that the class of augmented neighborhood models is modally equivalent to the class of relational models. The central idea is that, for each augmented set $\mathcal{X} \subseteq \wp(W)$, there is a relation $R_\mathcal{X} \subseteq W \times W$ defined as follows: $w \ R_\mathcal{X} \ v$ iff $v \in \bigcap \mathcal{X}$. The definition of $R_\mathcal{X}$ makes sense only if the collection of subsets $\mathcal{X}$ is augmented. In this section, I show that there are more subtle connections between neighborhood models (and non-normal modal logics) and relational models (and normal modal logics).

The first connection is that *every* collection of subsets $\mathcal{X} \subseteq \wp(W)$ can be associated with an ordering $\preceq_\mathcal{X} \subseteq W \times W$. For each $w, v \in W$, let $w \preceq_\mathcal{X} v$ iff for all $X \in \mathcal{X}$, if $w \in X$, then $v \in X$. That is, v is ranked at least as high as w by $\preceq_\mathcal{X}$ provided that every set from $\mathcal{X}$ containing w also contains v. This is a well-known definition in point-set topology (it is called the *specialization ordering*), the study of preferences (Andreka et al. 2002; Liu 2011) and multi-criteria decision-making (Dietrich and List 2013; de Jongh and Liu 2009). In Sect. 2.6.1, I use this ordering to facilitate a rigorous comparison between the evidence models from Sect. 1.4.4 and *plausibility models*, which are well known in the study of modal logics of belief and belief revision.

The second connection is based on neighborhood models themselves. On the face of it, the definition of truth of the modal operator on a neighborhood model seems to be a *second-order* statement since it asserts the existence of a subset implying either that a formula is true or that it is equal to the truth set of a formula. However, appearances are deceiving. By treating the neighborhoods as states in a larger model, there is a way to translate every neighborhood model into a relational model. Building on this idea, it can be shown that every non-normal modal logic can be *simulated* by a normal modal logic (Sect. 2.6.2), and that there is a translation of non-normal modal logic into first-order logic (Sect. 2.6.3).

2.6.1 From Neighborhoods to Orders

A **plausibility ordering** on a set of states W is a reflexive and transitive relation $\preceq \subseteq W \times W$. The intended meaning of $w \preceq v$ is that "(according to the agent) world v is at least as plausible as w". Plausibility models are widely used as a semantics for modal logics of belief (van Benthem 2004; Baltag and Smets 2006b, a; Girard 2008 and deontic logics Hansson (1990); van Benthem et al. (2014)).

Definition 2.54 (*Plausibility model*) A **plausibility model** is a tuple $\mathfrak{M} = \langle W, \preceq, V \rangle$ where W is a finite nonempty set; $\preceq \subseteq W \times W$ is a reflexive and transitive relation on W; and $V : \mathsf{At} \to \wp(W)$ is a valuation function. If $\preceq$ is also *connected* (for each $w, v \in W$, either $w \preceq v$ or $v \preceq w$), then we say that $\mathfrak{M}$ is a **connected plausibility model**. A pair $\mathfrak{M}, w$ where w is a state is called a **pointed (connected) plausibility model**.

When two worlds w and v cannot be compared by the plausibility ordering (for an agent), the interpretation is that the agent has either accepted contradictory evidence or lacks enough evidence to compare the two states.[20]

A number of different modal languages have been used to reason about plausibility structures. For instance, let $\mathcal{L}^{pl}(\mathsf{At})$, where At is the set of atomic propositions, be the smallest set of formulas generated by the following grammar:

$$p \mid \neg \varphi \mid \varphi \wedge \psi \mid [B]\psi \mid [\preceq]\varphi \mid [A]\varphi$$

where $p \in \mathsf{At}$. For each $\bigcirc \in \{B, \preceq, A\}$, let $\langle\bigcirc\rangle\varphi$ be defined as $\neg[\bigcirc]\neg\varphi$. Before defining truth for this language, I need some notation. For $w, v \in W$, write $w \prec v$ if $w \preceq v$ and $v \not\preceq w$. For $X \subseteq W$, let

$$Max_{\preceq}(X) = \{w \in X \mid \text{ there is no } v \in X \text{ such that } w \prec v\}.$$

For each set X, $Max_{\preceq}(X)$ is the set of most plausible worlds in X (i.e., the maximal elements of X according to the plausibility order $\preceq$).[21] Suppose that $\mathfrak{M} = \langle W, \preceq, V \rangle$ is a plausibility model with $w \in W$. Truth of the Boolean connectives and atomic propositions is defined as usual. I give only the clauses for the modal operators:

- $\mathfrak{M}, w \models [B]\varphi$ iff $Max_{\preceq}(W) \subseteq [\![\varphi]\!]_{\mathcal{M}}$
- $\mathfrak{M}, w \models [\preceq]\varphi$ iff for all $v \in W$, if $w \preceq v$ then $\mathcal{M}, v \models \varphi$
- $\mathfrak{M}, w \models [A]\varphi$ iff for all $v \in W$, $\mathcal{M}, v \models \varphi$.

So, φ is believed provided that φ is true throughout all of the most plausible states. There is much more to say about plausibility structures, their relationship with theories of belief revision, and the modal logic of beliefs (see van Benthem 2011 and Pacuit 2013a for discussions). In the remainder of this section, I focus on the relationship between plausibility models and neighborhood models.

From Plausibility Structures to Neighborhood Structures

There is a natural subset space associated with every plausibility model.

Definition 2.55 (*Upwards Closed Sets*) Suppose that $\preceq$ is a plausibility ordering on a set of states W (i.e., a reflexive and transitive relation on W). The upwards closure of a set X, denoted $\uparrow X$, is the set

$$\uparrow X = \{v \in W \mid \text{ there is a } w \in X \text{ such that } w \preceq v\}.$$

A set X is $\preceq$**-closed** when $\uparrow X \subseteq X$. Let $\mathcal{F}_{\preceq} = \{\uparrow X \mid X \subseteq W\}$ be the set of $\preceq$-closed sets.

---

[20] Swanson (2011) has an extensive discussion of incomparability when modeling conditionals.

[21] To keep things simple, I assume that the set of worlds is finite, so this maximal set always exists. One needs a (converse) well-foundedness condition to guarantee this when there are infinitely many states.

2.6 Translations

Exercise 2.28 If $\preceq$ is a plausibility ordering on W, then $\mathcal{F}_{\preceq}$ is closed under non-empty intersections. (Why do I need to specify non-empty intersections?) Is $\mathcal{F}_{\preceq}$ closed under supersets?

Using the above notation, there is a straightforward way to turn any plausibility model into a neighborhood structure. Let $\mathfrak{M} = \langle W, \preceq, V \rangle$ be a plausibility model. The associated neighborhood model is the model $\mathcal{M}^{\preceq} = \langle W^{\preceq}, N^{\preceq}, V^{\preceq} \rangle$, where $W^{\preceq} = W$; for each $w \in W$, $N^{\preceq}(w) = \mathcal{F}_{\preceq}$; and $V^{\preceq} = V$. Thus, the associated neighborhood model $\mathcal{M}^{\preceq}$ has a *uniform* neighborhood function (each state is associated with the same collection of sets).

Remark 2.56 A more general definition of plausibility models is possible in which each state is associated with a different plausibility ordering. That is, for each $w \in W$, $\preceq_w$ is a plausibility ordering on W. In this case, the neighborhoods may vary at each state: $N^{\preceq}(w) = \mathcal{F}_{\preceq_w}$. I focus on a single, global plausibility ordering to simplify the discussion.

The next step is to show that every plausibility model $\mathfrak{M} = \langle W, \preceq, V \rangle$ is *equivalent* to the corresponding neighborhood model $\mathcal{M}^{\preceq}$. Here, we face a problem that did not arise in the previous sections. The notion of equivalence between classes of models discussed in Sect. 2.1 assumes that there is a single underlying language that can be interpreted on *both* classes of models. However, I have not given a definition of truth for the language $\mathcal{L}^{pl}$ over neighborhood models.

One solution is to find a translation between an appropriate language interpreted over neighborhood models and $\mathcal{L}^{pl}$. Before defining such a translation, I note an important fact about the language $\mathcal{L}^{pl}$: We can restrict attention to formulas from $\mathcal{L}^{pl}$ that include only the $[\preceq]$ and $[A]$ modalities.

Fact 2.57 *On finite plausibility models, the belief modality* $[B]$ *is definable in terms of the* $[A]$ *and* $[\preceq]$ *modalities. That is,*

- $[B]\varphi \leftrightarrow [A]\langle \preceq \rangle [\preceq]\varphi$ *is valid on finite plausibility models.*[22]

The proof is straightforward given the following observation. The set of maximal elements in a plausibility model can be partitioned into *final clusters*:

Definition 2.58 (*Final Cluster*) Let $\mathfrak{M} = \langle W, \preceq, V \rangle$ be a plausibility model. A **final cluster** in $\mathcal{M}$ is a set $X \subseteq Max_{\preceq}(W)$ that is maximal and completely connected: for any $x, y \in X$, $x \preceq y$ and $y \preceq x$, and there is no $v \in W$ such that $w \prec v$ for some $w \in X$.

In a connected plausibility model, there is only one largest final cluster: the set $Max_{\preceq}(W)$. However, when the order is not connected, there may be disjoint final clusters. Using this terminology, (on finite models) $[B]\varphi$ is true provided that φ is true throughout all final clusters.

---

[22]This was first discussed by Boutilier (1992).

Exercise 2.29 Prove Fact 2.57.

Let $\mathcal{L}_0^{pl}$ be the sublanguage of $\mathcal{L}^{pl}$ without the belief modality $[B]$. The appropriate language for neighborhood models is $\mathcal{L}^{\langle\,],[A]}$—the propositional modal language generated by adding the neighborhood modality $\langle\,]$ and the universal modality $[A]$ to a propositional language. The definition of truth for formulas of the form $\langle\,]\varphi$ is given in Sect. 1.2.2 (Definition 1.2.2), and the clause for the universal modality $[A]$ is exactly as it is for the plausibility models given above. I can now define the translation between these two languages.

Definition 2.59 ($\preceq$-*translation*) The translation $tr_{\preceq} : \mathcal{L}^{\langle\,],[A]} \to \mathcal{L}_0^{pl}$ is defined by induction on formulas in $\mathcal{L}^{\langle\,],[A]}$:

- for each $p \in \mathsf{At}$, $tr_{\preceq}(p) = p$;
- $tr_{\preceq}(\neg\varphi) = \neg tr_{\preceq}(\varphi)$ and $tr_{\preceq}(\varphi \wedge \psi) = tr_{\preceq}(\varphi) \wedge tr_{\preceq}(\psi)$;
- $tr_{\preceq}([A]\varphi) = [A](tr_{\preceq}(\varphi))$; and
- $tr_{\preceq}(\langle\,]\varphi) = \langle A \rangle [\preceq](tr_{\preceq}(\varphi))$.

Proposition 2.13 *Let* $\mathfrak{M} = \langle W, \preceq, V \rangle$ *be a plausibility model. For any* $\varphi \in \mathcal{L}^{\langle\,],[A]}$ *and state* $w \in W$,

$$\mathfrak{M}, w \models tr_{\preceq}(\varphi) \text{ iff } \mathcal{M}^{\preceq}, w \models \varphi.$$

The proof is straightforward and left to the reader. However, this is a weak result. The conclusion is simply that every plausibility model "contains" a neighborhood model. Furthermore, anything that can be expressed in the language $\mathcal{L}^{\langle\,],[A]}$ can be translated into the language $\mathcal{L}^{pl}$. Of course, this translation is not surjective. That is, there are formulas of $\mathcal{L}^{pl}$ that are not the translation of some formula from $\mathcal{L}^{\langle\,],[A]}$. So, we cannot conclude that for every plausibility model, there is a *modally equivalent* neighborhood model (at least with respect to the language $\mathcal{L}^{pl}$).

From Neighborhood Models to Plausibility Models

There is also a natural way to define a plausibility ordering given any subset space. The approach is to use the so-called *specialization order*, a notion that occurs in point-set topology (cf. Sect. 1.4.4) and in recent theories of relation merge (cf. Andreka et al. 2002; Liu 2011).

Definition 2.60 (*Specialization Order*) Suppose that $\langle W, \mathcal{F} \rangle$ is a subset space. Define $\preceq_{\mathcal{F}} \subseteq W \times W$ as follows:

$$w \preceq_{\mathcal{F}} v \text{ iff for all } X \in \mathcal{F}, \text{ if } w \in X, \text{ then } v \in X.$$

The intuition is that v is "at least as special" as w provided that every set in $\mathcal{F}$ that contains w also contains v. If $\mathcal{F}$ is a set of evidence as in the evidence models from Sect. 1.4.4, then $w \preceq_{\mathcal{F}} v$ means that every piece of evidence that supports w (i.e., contains w) also supports v, though there might be some pieces of evidence that

2.6 Translations

support v but not w. To make this definition a bit more concrete, here is a simple illustration.

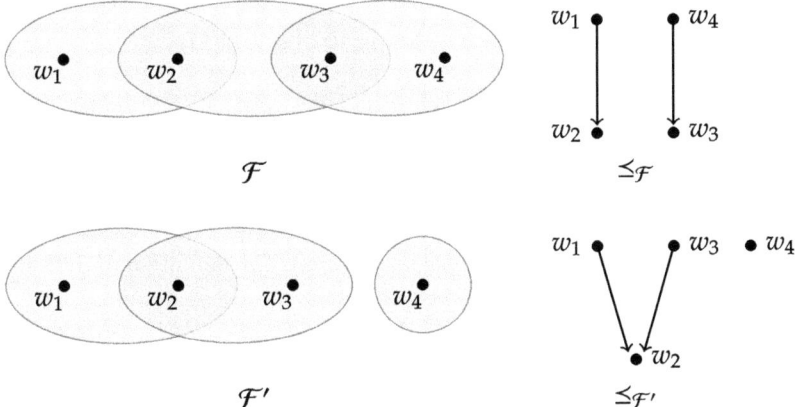

Of course, the relational properties of $\preceq_{\mathcal{F}}$ depend on the algebraic properties of $\mathcal{F}$. However, all specialization orders are reflexive and transitive:

Observation 2.61 *Suppose that $\langle W, \mathcal{F} \rangle$ is a subset space. Then, $\preceq_{\mathcal{F}}$ is a transitive and reflexive relation on W.*

Proof Suppose that $w \preceq_{\mathcal{F}} v$ and $v \preceq_{\mathcal{F}} y$. Suppose, also, that $X \in \mathcal{F}$ and $w \in X$. Then, since $w \preceq_{\mathcal{F}} v$, we have $v \in X$. Since, $v \preceq_{\mathcal{F}} y$, we have $y \in X$. Thus, $w \preceq_{\mathcal{F}} y$. Clearly, $\preceq_{\mathcal{F}}$ is reflexive. □

The examples given above show that, in general, the specialization order $\preceq_{\mathcal{F}}$ is not connected.

Exercise 2.30   1. What property of $\mathcal{F}$ guarantees that the specialization ordering $\preceq_{\mathcal{F}}$ is connected? Recall that a relation $R \subseteq X \times X$ is **connected** if every pair of distinct elements from X is related. That is, for all $x, y \in X$, if $x \neq y$, then either $x \mathrel{R} y$ or $y \mathrel{R} x$.
2. Suppose that $\preceq$ is the reflexive, transitive closure of $\{(w, v), (v, x), (w, y), (y, x)\}$ (that is, $\preceq$ is the smallest relation containing this set that is reflexive and transitive). Find a subset space $\mathcal{F}$ on $W = \{w, v, y, x\}$ such that $\preceq \, = \, \preceq_{\mathcal{F}}$.

There are (at least) two different ways to associate a plausibility model with a neighborhood structure $\mathcal{M} = \langle W, N, V \rangle$. The first is to assign to each state $w \in W$, a plausibility ordering $\preceq_{N(w)}$. Strictly speaking, unless N is a constant function, this is not a plausibility model according to Definition 2.54 since each state is assigned a different plausibility ordering (cf. Remark 2.56). Rather than pursuing this line of thought, I focus on the relationship between plausibility models and the evidence models from Sect. 1.4.4. It turns out that the precise relationship is subtle. Following the discussions in van Benthem et al. (2014) and van Benthem and Pacuit (2011, Sect. 5), I briefly discuss this relationship in the remainder of this section, focusing on uniform evidence models.

The first step is to combine the two languages $\mathcal{L}^{ev}$ and $\mathcal{L}^{pl}$. Let $\mathcal{L}^{epl}(\mathsf{At})$ be the smallest set of formulas generated by the following grammar:

$$p \mid \neg\varphi \mid (\varphi \wedge \psi) \mid \langle\,]\varphi \mid [\preceq]\varphi \mid [B]\varphi \mid [A]\varphi$$

where $p \in \mathsf{At}$. A model for this language includes relations for the $[\preceq]$ and $[B]$ modalities (cf. Appendix A) and a neighborhood function for the $\langle\,]$ modality ($[A]$ is the universal modality). Given such a model, the definition of truth for formulas from $\mathcal{L}^{epl}$ follows the usual pattern (cf. Definition A.3). For example, suppose that $\mathcal{M} = \langle W, E, \preceq, B, V \rangle$, where E is a neighborhood function, $B \subseteq W \times W$, $\preceq \,\subseteq W \times W$ and $V : \mathsf{At} \to \wp(W)$ is a valuation function. The definition of truth for the modalities at $w \in W$ is:

- $\mathcal{M}, w \models \langle\,]\varphi$ iff there is an $X \in E(w)$ such that for all $v \in X$, $\mathcal{M}, v \models \varphi$.
- $\mathcal{M}, w \models [\preceq]\varphi$ iff for all $v \in W$, if $w \preceq v$, then $\mathcal{M}, v \models \varphi$.
- $\mathcal{M}, w \models [B]\varphi$ iff for all $v \in W$, if $w \, B \, v$, then $\mathcal{M}, v \models \varphi$.
- $\mathcal{M}, w \models [A]\varphi$ iff for all $v \in W$, $\mathcal{M}, v \models \varphi$.

Given the intended interpretation of the modalities in $\mathcal{L}^{epl}$, it is natural to impose the following constraints on a model $\langle W, E, \preceq, B, V \rangle$:

1. for each $w \in W$, $\emptyset \notin E(w)$ and $W \in E(w)$;
2. for all $w, v, u \in W$, if $w \preceq v$ and $w \in X \in E(u)$, then $v \in X$; and
3. if $w \preceq v$ and $u \, B \, w$, then $u \, B \, v$.

To see the importance of these constraints, consider a model $\mathcal{M} = \langle W, \mathcal{E}, \preceq_\mathcal{E}, B, V \rangle$, where $\mathcal{E}$ is a uniform evidence function, $\preceq_\mathcal{E}$ is the specialization order defined from $\mathcal{E}$ and $B \subseteq W \times W$. It is easy to find such a model with states w, v and u such that $w \, B \, v$ and $v \preceq_\mathcal{E} u$; yet it is not the case that $w \, B \, u$. For example, let $W = \{w, v\}$ and $\mathcal{E} = \{W\}$, and $B = \{(w, w)\}$. Then, $w \, B \, w$ and $w \preceq_\mathcal{E} v$; yet it is not the case that $w \, B \, v$. However, this does not happen on intended models:

Definition 2.62 (*Transforming Evidence Models*) Let $\mathcal{M} = \langle W, \mathcal{E}, V \rangle$ be a uniform evidence model (recall Definition 1.35). An **extended evidence model** is a structure $\mathcal{M}^\Delta = \langle W, \mathcal{E}, B_\mathcal{E}, \preceq_\mathcal{E}, V \rangle$, where

- $w \, B_\mathcal{E} \, v$ iff $v \in \bigcap \mathcal{X}$ for some scenario $\mathcal{X}$ from $\mathcal{E}$; and
- $w \preceq_\mathcal{E} v$ iff for any $X \in \mathcal{E}$, if $w \in X$, then $v \in X$.

The following exercise illustrates key properties satisfied by extended evidence models.

Exercise 2.31 Suppose that $\mathcal{M} = \langle W, \mathcal{E}, V \rangle$ is a uniform evidence model, and let $\preceq_\mathcal{E}$ be defined as above. For each $w \in W$, let $\mathcal{E}[w] = \{X \in \mathcal{E} \mid w \in X\}$. Prove the following two statements (cf. van Benthem et al. 2014, Lemma 4).

1. For each $w \in W$, if $w \in \bigcap \mathcal{X}$ for some scenario $\mathcal{X} \subseteq \mathcal{E}$, then w is $\preceq_\mathcal{E}$-maximal. Furthermore, if $\mathcal{X}$ is a non-empty scenario from $\mathcal{E}$, then $\mathcal{X} = \mathcal{E}[w]$ for $\preceq_\mathcal{E}$-maximal state w.

2. Suppose that $\mathcal{M}$ is also flat. For each $w \in W$, if w is $\preceq_{\mathcal{E}}$-maximal, then $w \in \bigcap \mathcal{X}$ for some scenario $\mathcal{X} \subseteq \mathcal{E}$. Furthermore, if w is $\preceq_{\mathcal{E}}$ maximal, then $\mathcal{E}[w]$ is a scenario.
3. Using items 1 and 2 to prove that $[A]\langle\preceq\rangle[\preceq]\varphi \to [B]\varphi$ is valid on extended uniform evidence models (that is, if $\mathcal{M} = \langle W, \mathcal{E}, V\rangle$ is a uniform evidence model, then the formula is valid on $\mathcal{M}^{\Delta}$). Furthermore, over the class of models that are, moreover, flat, the formula $[A]\langle\preceq\rangle[\preceq]\varphi \leftrightarrow [B]\varphi$ is valid (cf. Fact 2.57).

Consult van Benthem et al. (2014) for the precise relationship between evidence models, extended evidence models and plausibility models.

2.6.2 The Normal Translation

In this section, I explore a deeper connection between neighborhood models and relational models. The key observation is that neighborhood models are just a special type of relational models.

The states in the these special relational models are divided into two sorts: the elements of a non-empty set W and the subsets from W. For any set W, let $W^{\circ} = W \cup \wp(W)$. (It is also important to assume that $W \cap \wp(W) = \emptyset$.) There are two natural relations on W° corresponding to the "element of" and "not element of" relations between states and subsets. It will be convenient to use the converse of the "(not) element of" relation:

- $R_{\ni} \subseteq \wp(W) \times W$ with $R_{\ni} = \{(U, w) \mid w \in W,\ U \in \wp(W),\ w \in U\}$.
- $R_{\not\ni} \subseteq \wp(W) \times W$ with $R_{\not\ni} = \{(U, w) \mid w \in W,\ U \in \wp(W),\ w \notin U\}$.

Two remarks about the above relations are in order. First, $R_{\not\ni}$ is *not* the complement of $R_{\ni}$ (with respect to $W^{\circ} \times W^{\circ}$). That is $R_{\not\ni} \neq (W^{\circ} \times W^{\circ}) - R_{\ni}$. This is, because, for example, $(w, v) \in (W^{\circ} \times W^{\circ}) - R_{\ni}$, where $w, v \in W$, but $(w, v) \notin R_{\not\ni}$. Of course, it is true that $R_{\not\ni} = (\wp(W) \times W) - R_{\ni}$.

Second, there are other relations that can be studied in this context. For instance, one can include the subset relation $R_{\subseteq} \subseteq \wp(W) \times \wp(W)$ with $R_{\subseteq} = \{(U, V) \mid U, V \in \wp(W),\ U \subseteq V\}$ in the model. A *relational-neighborhood* model includes a third relation between states W and $\wp(W)$ (which is intended to represent a neighborhood function).

Definition 2.63 (*Relational-Neighborhood Model*) Suppose that W is a non-empty set of states and At is a set of atomic propositions. A **relational neighborhood model** on W is a tuple $\langle W^{\circ}, R_N, R_{\ni}, R_{\not\ni}, V\rangle$, where $W^{\circ} = W \cup \wp(W)$, $R \subseteq W \times \wp(W)$, $R_{\ni} \subseteq \wp(W) \times W$ with $R_{\ni} = \{(U, w) \mid w \in W,\ U \in \wp(W),\ w \in U\}$, $R_{\not\ni} \subseteq \wp(W) \times W$ with $R_{\not\ni} = \{(U, w) \mid w \in W,\ U \in \wp(W),\ w \notin U\}$, and $V : \text{At} \to \wp(W)$ is a valuation function.

The definition of the valuation function in a relational-neighborhood function highlights the two-sorted aspect of these models. Since the range of the valuation function is the set of states W, the atomic propositions (and, hence, all non-modal formulas) are "state formulas" that can express properties of the set of states W.[23] This also means that atomic propositions are interpreted as false at all subsets of W.[24] A more standard approach would be to let $V : \mathsf{At} \to \wp(W^\circ)$, allowing atomic propositions to be interpreted at both states and subsets.

There is a natural modal language associated with the above models. Let $\mathcal{L}_2$ be the smallest set of formulas generated by the following grammar:

$$p \mid \neg\varphi \mid \varphi \wedge \psi \mid [\ni]\varphi \mid [\not\ni]\varphi \mid [R]\varphi$$

where $p \in \mathsf{At}$. Suppose that $\mathfrak{M} = \langle W^\circ, R_\ni, R_{\not\ni}, R, V \rangle$ is a relational-neighborhood model with $W^\circ = W \cup \wp(W)$. Truth of formulas $\varphi \in \mathcal{L}_2$ at elements $x \in W^\circ$ is defined as follows:

- $\mathfrak{M}, x \models p$ iff $x \in W$ and $x \in V(p)$.
- $\mathfrak{M}, x \models \neg\varphi$ iff $x \in W$ and $\mathcal{M}, x \not\models \varphi$.
- $\mathfrak{M}, x \models \varphi \wedge \psi$ iff $x \in W$, $\mathcal{M}, x \models \varphi$, and $\mathcal{M}, x \models \psi$.
- $\mathfrak{M}, x \models [R]\varphi$ iff $x \in W$ and for all $y \in W^\circ$, if $u \, R \, y$, then $\mathcal{M}, y \models \varphi$.
- $\mathfrak{M}, x \models [\ni]\varphi$ iff $x \in \wp(W)$ and for all $y \in W^\circ$, if $x \, R_\ni \, y$, then $\mathcal{M}, y \models \varphi$.
- $\mathfrak{M}, x \models [\not\ni]\varphi$ iff $x \in \wp(W)$ for all $y \in W^\circ$, if $x \, R_{\not\ni} \, y$, then $\mathcal{M}, y \models \varphi$.

According to the above definition, subsets of W behave similarly to impossible worlds (cf. Sect. 2.2.4) since all non-modal formulas and formulas of the form $[R]\varphi$ are false at these points. This means that, for instance, $p \vee \neg p$ is false at points $X \in \wp(W)$. In order to express the two-sorted nature of the states in a relational-neighborhood model, it is sometimes convenient to include a special proposition St with the fixed interpretation $V(\mathsf{St}) = W$. Given the above remark, if $p \in \mathsf{At}$, then we can *define* St to be $p \vee \neg p$.

It is not hard to see that every neighborhood model can be transformed into a relational-neighborhood model. Suppose that $\mathcal{M} = \langle W, N, V \rangle$ is a neighborhood model. The associated relational-neighborhood model is $\mathcal{M}^\circ = \langle W^\circ, R_\ni, R_{\not\ni}, R_N, V^\circ \rangle$, where, for all $p \in \mathsf{At}$, $V^\circ(p) = V(p)$ and

- $R_N = \{(w, U) \mid w \in W, U \in \wp(W), U \in N(w)\}$.

To simplify the notation, I write '$[N]\varphi$' instead of '$[R_N]\varphi$'.

The main observation of this section is that there is a translation of the basic modal language $\mathcal{L}$ into $\mathcal{L}_2$ that preserves truth.

Definition 2.64 (*Normal Translation*) The **normal translation** of the basic modal language $\mathcal{L}$ is a function $\mathsf{nt} : \mathcal{L} \to \mathcal{L}_2$ defined by induction on the structure of $\varphi \in \mathcal{L}$:

- $\mathsf{nt}(p) = p$.
- $\mathsf{nt}(\neg\varphi) = \neg\mathsf{nt}(\varphi)$.

---

[23] This means that atomic propositions are *nominals* (cf. Sect. 3.1 and Areces and ten Cate 2007) with respect to elements in the domain of the model that correspond to subsets.

[24] Another option would be to let atomic propositions be *undefined* at all $U \in \wp(W)$.

2.6 Translations

- $\mathsf{nt}(\varphi \wedge \psi) = \mathsf{nt}(\varphi) \wedge \mathsf{nt}(\varphi)$.
- $\mathsf{nt}(\Box \varphi) = \langle N \rangle ([\ni]\mathsf{nt}(\varphi) \wedge [\not\ni]\neg\mathsf{nt}(\varphi))$.

The next proposition shows that the translation works as expected.

Proposition 2.14 *Suppose that* $\mathcal{M} = \langle W, N, V \rangle$ *is a neighborhood model. For all* $\varphi \in \mathcal{L}$, *for all* $w \in W$,

$$\mathcal{M}, w \models \varphi \text{ iff } \mathcal{M}^\circ, w \models \mathsf{nt}(\varphi).$$

Proof The proof is by induction on $\varphi \in \mathcal{L}$. The proof for the base case and the boolean connectives is straightforward. I give the proof only for the case in which φ is of the form $\Box \psi$. Note that the induction hypothesis is as follows:

IH for all $w \in W$, $\mathcal{M}, w \models \psi$ iff $\mathcal{M}^\circ, w \models \mathsf{nt}(\psi)$.

Suppose that $w \in W$ with $\mathcal{M}, w \models \Box \psi$. We must show that $\mathcal{M}^\circ, w \models \mathsf{nt}(\Box \psi)$. That is, we must show that $\mathcal{M}^\circ, w \models \langle N \rangle ([\ni]\mathsf{nt}(\psi) \wedge [\not\ni]\neg\mathsf{nt}(\psi))$. Suppose that $X = [\![\psi]\!]_{\mathcal{M}}$. By the construction of $\mathcal{M}^\circ$, since $[\![\psi]\!]_{\mathcal{M}} \in N(w)$, we have that $w\ R_N\ X$. Thus, we must show that $\mathcal{M}^\circ, X \models [\ni]\mathsf{nt}(\psi) \wedge [\not\ni]\mathsf{nt}(\psi)$. The induction hypothesis implies that

$(*)$ for all $v \in W$, $v \in [\![\psi]\!]_{\mathcal{M}}$ iff $\mathcal{M}^\circ, v \models \mathsf{nt}(\psi)$.

By the construction of $\mathcal{M}^\circ$, we have for all $x \in W \cup \wp(W)$, $X\ R_\ni\ x$ iff $x \in W$ and $x \in X$. Furthermore, by $(*)$, for all $v \in W$, if $v \in X$, then $\mathcal{M}^\circ, v \models \mathsf{nt}(\psi)$. This implies that $\mathcal{M}^\circ, X \models [\ni]\mathsf{nt}(\psi)$. A similar argument shows that $\mathcal{M}^\circ, X \models [\not\ni]\neg\mathsf{nt}(\psi)$. Thus, $\mathcal{M}^\circ, w \models \langle N \rangle ([\ni]\mathsf{nt}(\psi) \wedge [\not\ni]\neg\mathsf{nt}(\psi))$, as desired.

Suppose that $\mathcal{M}^\circ, w \models \mathsf{nt}(\Box \psi)$. That is, suppose that

$$\mathcal{M}^\circ, w \models \langle N \rangle ([\ni]\mathsf{nt}(\psi) \wedge [\not\ni]\neg\mathsf{nt}(\psi)).$$

Then, there is some $x \in W \cup \wp(W)$ such that $w\ R_N\ x$ and

$$\mathcal{M}^\circ, x \models [\ni]\mathsf{nt}(\psi) \wedge [\not\ni]\neg\mathsf{nt}(\psi).$$

By construction of $\mathcal{M}^\circ$, since $w\ R_N\ x$, $x = X \subseteq W$ with $X \in N(w)$. Thus, in order to show that $\mathcal{M}, w \models \Box \psi$, we must show that $X = [\![\psi]\!]_{\mathcal{M}}$. Suppose that $v \in X$. By the definition of $R_\ni$, we have $X\ R_\ni\ v$ (and $v \in W$). Since $\mathcal{M}^\circ, X \models [\ni]\mathsf{nt}(\psi)$, we have $\mathcal{M}^\circ, v \models \mathsf{nt}(\psi)$. By the induction hypothesis, $\mathcal{M}, v \models \psi$. Hence, $X \subseteq [\![\psi]\!]_{\mathcal{M}}$. Conversely, suppose that $v \notin X$. Then, by the definition of $R_{\not\ni}$, we have $X\ R_{\not\ni}\ v$ (and $v \in W$). Since $\mathcal{M}^\circ, X \models [\not\ni]\neg\mathsf{nt}(\psi)$, we have $\mathcal{M}^\circ, v \models \neg\mathsf{nt}(\psi)$. Since $\mathcal{M}^\circ, v \not\models \mathsf{nt}(\psi)$ and $v \in W$, by the induction hypothesis, $\mathcal{M}, v \not\models \psi$. Hence, $[\![\psi]\!]_{\mathcal{M}} \subseteq X$. Therefore, since $X \in N(w)$ and $X = [\![\psi]\!]_{\mathcal{M}}$, we have that $\mathcal{M}, w \models \Box \psi$, as desired. $\square$

To illustrate the above translation, let $\mathcal{M} = \langle W, N, V \rangle$ be a neighborhood model with $W = \{w, v\}$, $N(w) = \{\{w\}, \{v\}\}$, $N(v) = \{\emptyset\}$, and $V(p) = \{w\}$. The relational model $\mathcal{M}^\circ$ is given below (the solid arrows correspond to the R_N relation; the dashed arrows correspond to the $R_\ni$ relation; and the dotted arrows correspond to the $R_{\not\ni}$ relation). To simplify the comparison between the models, I also draw the neighborhood model $\mathcal{M}$.

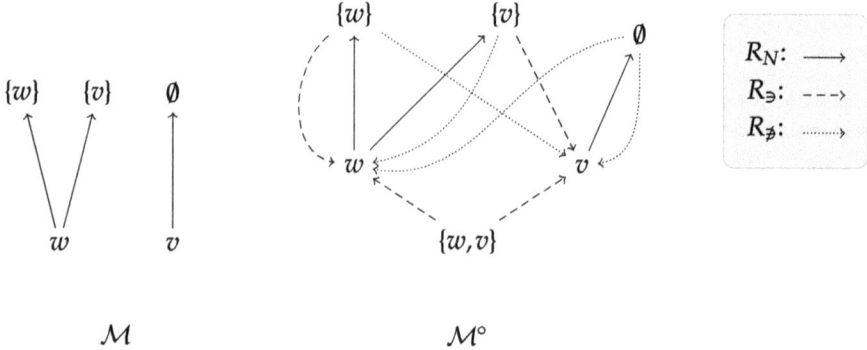

Note that $\mathcal{M}, w \models \Box p$ and $\mathcal{M}, v \models \Box \bot$. As the reader is invited to check, the following is statements are true:

- $\mathcal{M}^\circ, w \models \langle N \rangle([\ni]p \wedge [\not\ni]\neg p)$ and $\mathcal{M}^\circ, v \not\models \langle N \rangle([\ni]p \wedge [\not\ni]\neg p)$; and
- $\mathcal{M}^\circ, v \models \langle N \rangle([\ni]\bot \wedge [\not\ni]\top)$ and $\mathcal{M}^\circ, w \not\models \langle N \rangle([\ni]\bot \wedge [\not\ni]\top)$.

Exercise 2.32 Find normal translations of: $\Box(p \vee \neg p)$, $\Box p \vee \Box \neg p$, and $\neg \Box p \vee \Box p$.

The normal translation can be simplified when the neighborhood models are closed under supersets.

Definition 2.65 (*Monotonic translation*) The **monotonic translation** of the basic modal language $\mathcal{L}$ is a function $\mathsf{mt} : \mathcal{L} \to \mathcal{L}_2$ defined by induction on $\mathcal{L}$, where the base case and Boolean connectives are as in Definition 2.64, and the modal clause is:

$$\mathsf{mt}(\Box \varphi) = \langle N \rangle [\ni] \mathsf{mt}(\varphi).$$

Proposition 2.15 *Suppose that* $\mathcal{M} = \langle W, N, V \rangle$ *is a monotonic neighborhood model. For all* $\varphi \in \mathcal{L}$, *for all* $w \in W$,

$$\mathcal{M}, w \models \varphi \text{ iff } \mathcal{M}^\circ, w \models \mathsf{mt}(\varphi)$$

Exercise 2.33 Prove Proposition 2.15 (it is similar to the proof of Proposition 2.14).

Kracht and Wolter (1999) use the normal and monotonic translations to show that all non-normal modal logics can be *simulated* by a (two-sorted) normal modal logic (cf. Gasquet and Herzig 1996).

Exercise 2.34 (Kracht and Wolter 1999) Use the fact the satisfiability problem for multi-modal normal modal logic is decidable and the above translation to prove that the satisfiability problem for the non-normal modal logic **E** is decidable. What can be concluded about the complexity of the satisfiability problem for **E**?

Exercise 2.35 Suppose that $\mathcal{M}_1$ and $\mathcal{M}_2$ are monotonically-bisimilar neighborhood models (Definition 2.2). Does this imply that $\mathcal{M}_1^\circ$ and $\mathcal{M}_2^\circ$ are relationally-bisimilar (cf. Definition A.9)? Is the converse true (i.e., if $\mathcal{M}_1^\circ$ and $\mathcal{M}_2^\circ$ are relationally-bisimilar, then $\mathcal{M}_1$ and $\mathcal{M}_2$ are monotonically-bisimilar)? Answer the same questions for neighborhood models that are not monotonic.

2.6.3 The Standard Translation

Building on the normal translation from the previous section and the well-known *standard translation* of normal modal logic into first-order logic, in this section, I show that non-normal modal logic can be viewed as a fragment of *first-order logic*. In the remainder of this section, I assume that the reader is familiar with first-order logic (specifically, I assume familiarity with the syntax and semantics of first-order logic, the notion of an *isomorphism* and basic model-theoretic results). Consult Enderton (2001) for the necessary background.

I start by defining a first-order language that can be interpreted on relational-neighborhood models (Definition 2.63). It will be convenient to work with a *two-sorted* first-order language. Formally, there are two sorts, $\{w, n\}$. Terms of the first sort (w) are intended to represent states, whereas terms of the second sort (n) are intended to represent neighborhoods (i.e., subsets of sort w). I assume that there are countable sets of variables of each sort. To simplify notation, I use the following conventions: $x, y, x', y', x_1, y_2, \ldots$ denote variables of sort w (*world variables*), and $u, v, u', v', u_1, v_1, \ldots$ denote variables of sort n (*neighborhood variables*).

The language is built from a signature containing a unary predicate P_i (of sort w) for each $i \in \mathbb{N}$, a binary relation symbol N relating elements of sort w to elements of sort n, and a binary relation symbol E relating elements of sort n to elements of sort w. The intended interpretation of xNu is "u is a neighborhood of x", and the intended interpretation of uEx is "x is an element of u". The language $\mathcal{L}_{\text{fo}}$ is built from the following grammar:

$$x = y \mid u = v \mid P_i x \mid xNu \mid uEx \mid \neg\varphi \mid \varphi \wedge \psi \mid \exists x\varphi \mid \exists u\varphi$$

where $i \in \mathbb{N}$; x and y are state variables; and u and v are neighborhood variables. The usual abbreviations (e.g. $\forall$ for $\neg\exists\neg$) apply. Note that $\exists x\varphi$ means that "*there is an element of sort w that satisfies φ*" (similarly, $\exists u\varphi$ means that "*there is an element of sort n that satisfies φ*").

Formulas of $\mathcal{L}_{\text{fo}}$ are interpreted in two-sorted first-order structures $\mathscr{M} = \langle D, \{P_i \mid i \in \mathbb{N}\}, R, E\rangle$, where $D = D^w \cup D^n$ (and $D^w \cap D^n = \emptyset$), each $P_i \subseteq D^w$, $R \subseteq D^w \times$

D^n and $E \subseteq D^n \times D^w$. The relation R is the interpretation of the binary relational symbol N.[25] As usual, the equality symbol is always interpreted as equality on the appropriate domain. In addition, the usual definitions of free and bound variables apply. Truth of sentences (formulas with no free variables) $\varphi \in \mathcal{L}_{\text{fo}}$ in a structure $\mathcal{M}$ (denoted $\mathcal{M} \Vdash \varphi$) is defined as expected. If x is a free state variable in φ (denoted $\varphi(x)$), then $\mathcal{M} \Vdash \varphi[w]$ means that φ is true in $\mathcal{M}$ when $w \in D^w$ is assigned to x. Note that $\mathcal{M} \Vdash \exists x \varphi$ iff there is an element $w \in D^w$ such that $\mathcal{M} \models \varphi[w]$. If Γ is a set of $\mathcal{L}_{\text{fo}}$-formulas, and $\mathcal{M}$ is an $\mathcal{L}_{\text{fo}}$-model, then $\mathcal{M} \Vdash \Gamma$ means that for all $\gamma \in \Gamma$, $\mathcal{M} \Vdash \gamma$. Given a class K of $\mathcal{L}_{\text{fo}}$-models, the *semantic consequence relation* over K is denoted $\Vdash_\mathsf{K}$. That is, for a set of $\mathcal{L}_{\text{fo}}$-formulas $\Gamma \cup \{\varphi\}$, $\Gamma \Vdash_\mathsf{K} \varphi$, if for all $\mathcal{M} \in \mathsf{K}$, $\mathcal{M} \Vdash \Gamma$ implies that $\mathcal{M} \Vdash \varphi$.

The basic modal language and neighborhood models can be translated into the first-order setting as follows.

Definition 2.66 (*First-Order Translations of Neighborhood Models*) Suppose that $\mathcal{M} = \langle W, N, V \rangle$ is a neighborhood model. The **first-order translation** of $\mathcal{M}$ is the structure $\mathcal{M}^\bullet = \langle D, \{P_i \mid i \in \mathbb{N}\}, R_N, R_\ni \rangle$, where

- $D = D^w \cup D^n$ with $D^w = W$, $D^n = N[W] = \bigcup_{w \in W} N(w)$.
- $P_i = V(p_i)$ for each $p_i \in \mathsf{At}$.
- $R_N = \{(w, X) \mid w \in D^w, X \in N(w)\}$.
- $R_\ni = \{(X, w) \mid w \in D^w, w \in X\}$.

Definition 2.67 (*Standard Translation*) The **standard translation** of the basic modal language $\mathcal{L}(\mathsf{At})$ (where $\mathsf{At} = \{p_i \mid i \in \mathbb{N}\}$ is a countable set of atomic propositional variables) is a family of functions $\mathsf{st}_x : \mathcal{L}(\mathsf{At}) \to \mathcal{L}_{\text{fo}}$ defined as follows: $\mathsf{st}_x(p_i) = P_i x$, $\mathsf{st}_x(\neg \varphi) = \neg \mathsf{st}_x(\varphi)$, $\mathsf{st}_x(\varphi \wedge \psi) = \mathsf{st}_x(\varphi) \wedge \mathsf{st}_x(\psi)$, and

$$\mathsf{st}_x(\Box \varphi) = \exists u (x N u \wedge (\forall y (u E y \leftrightarrow \mathsf{st}_y(\varphi))))$$

Exercise 2.36 Prove that the standard translation preserves truth. That is, prove the following lemma.

Lemma 2.24 *Let $\mathfrak{M} = \langle W, N, V \rangle$ be a neighborhood model and $\varphi \in \mathcal{L}$. For each $w \in W$, $\mathcal{M}, w \models \varphi$ iff $\mathcal{M}^\bullet \models \mathsf{st}_x(\varphi)[w]$.*

So, every neighborhood model can be associated with an $\mathcal{L}_{\text{fo}}$-model that preserves truth of the basic modal language (using the standard translation). However, it is not the case that every $\mathcal{L}_{\text{fo}}$-structure is the translation of a neighborhood model.[26] Fortunately, it is possible to axiomatize the class of translations of neighborhood models. Let $\mathsf{N} = \{\mathcal{M} \mid \mathcal{M} \cong \mathcal{M}^\bullet$ for some neighborhood model $\mathcal{M}\}$, where $\mathcal{M} \cong \mathcal{N}$ means that there is an **isomorphism**[27] between $\mathcal{M}$ and $\mathcal{N}$, and let **NAX** be the following axioms:

---

[25]I am not using 'N' since that is used to denote neighborhood functions.

[26]This should be contrasted with the standard translation of relational models for normal modal logic. Consult Blackburn et al. (2001), Sect. 2.4, for details.

[27]In this context, an isomorphism between $\mathcal{L}_{\text{fo}}$-models $\mathcal{M} = \langle D, \{P_i \mid i \in \mathbb{N}\}, R, E \rangle$ and $\mathcal{M}' = \langle D', \{P'_i \mid i \in \mathbb{N}\}, R', E' \rangle$ is a 1-1 and onto function $f : D \to D'$ satisfying the structural conditions: $w \in P_i$ iff $f(w) \in P'_i$, $w R u$ iff $f(w) R' f(u)$ and $u E w$ iff $f(u) E' f(w)$.

2.6 Translations

(A1) $\exists x(x = x)$
(A2) $\forall u \exists x (x N u)$
(A3) $\forall u \forall v (\neg(u = v) \rightarrow \exists x((uEx \wedge \neg vEx) \vee (\neg uEx \wedge vEx)))$

It is not hard to see that if $\mathcal{M}$ is a neighborhood model, then $\mathcal{M}^\bullet \Vdash \mathbf{NAX}$. The next result states that **NAX** completely characterizes the class N.

Proposition 2.16 *Suppose $\mathcal{M}$ is an $\mathcal{L}_{\text{fo}}$-model and $\mathcal{M} \Vdash \mathbf{NAX}$. Then, there is a neighborhood model $\mathcal{M}_\bullet$ such that $\mathcal{M} \cong (\mathcal{M}_\bullet)^\bullet$.*

Proof Let $\mathcal{M} = \langle D^w \cup D^n, \{P_i \mid i \in \omega\}, R, E \rangle$ be an $\mathcal{L}_{\text{fo}}$-model such that $\mathcal{M} \Vdash \mathbf{NAX}$. We will construct a neighborhood model $\mathcal{M}_\bullet = \langle W, N, V \rangle$ such that $\mathcal{M} \cong (\mathcal{M}_\bullet)^\bullet$. First, define a map $\nu : D^n \to \wp(D^w)$ by $\nu(u) = \{w \in D^w \mid uEw\}$ and let $W = D^w$. Since $\mathcal{M} \Vdash$ (A1), $W \neq \emptyset$. Now define for each $w \in W$ and each $X \subseteq W$: $X \in N(w)$ iff there is a $u \in D^n$ such that sNu and $X = \nu(u)$, and define for all $i \in \mathbb{N}$, $V(p_i) = \{w \in W \mid \mathcal{M} \models P_i[w]\}$. Then, $\mathcal{M}_\bullet$ is clearly a well-defined neighborhood model. The proof is concluded if we can show that the map

$$f : D^w \cup D^n \to W \cup \bigcup_{w \in W} N(w),$$

defined as $f(w) = w$ for $w \in D^w$ and $f(u) = \nu(u)$ for $u \in D^n$, is an isomorphism from $\mathcal{M}$ to $(\mathcal{M}_\bullet)^\bullet = \langle W \cup N[W], \{P'_i \mid i \in \omega\}, R_N, R_\ni \rangle$ (cf. Definition 2.66).

First, it follows directly from $\mathcal{M} \Vdash$ (A3), that ν is injective. Second, by the definition of ν and the $(\cdot)^\bullet$-construction, the range of ν, denoted $\mathsf{rng}(\nu)$, contains $\bigcup_{w \in D^w} \nu(w) = \bigcup_{w \in W} N(w)$. The inclusion $\mathsf{rng}(\nu) \subseteq \bigcup_{w \in W} N(w)$ follows from the assumption that $\mathcal{M} \Vdash$ (A2) since this implies that, for every $u \in D^n$, there is a $w \in D^w$ such that $\nu(u) \in N(w)$. The structural conditions follow directly by construction: For all $i \in \mathbb{N}$, $w \in P_i$ iff $w \in V(p_i)$ iff $w \in P'_i$. Similarly, for all $w \in D^w$ and all $u \in D^n$: $w R u$ iff $\nu(u) \in N(w)$ iff $w R_N \nu(u)$, and $u E w$ iff $w \in \nu(u)$ iff $\nu(u) R_\ni w$. □

Thus, in a precise way, models in N can be viewed as neighborhood models. This means that (monotonic) bisimulations, bounded morphisms, disjoint unions, and other model constructions on neighborhood models (cf. Sect. 2.1) can be applied to models in N. For instance, an $\mathcal{L}_{\text{fo}}$-formula $\alpha(x)$ is said to be **invariant under behavioral equivalence** (Definition 2.9) on the class N iff for all w-elements w from $\mathcal{M}$ and w-elements v from $\mathcal{N}$, if $\mathcal{M}_\bullet, w$ and $\mathcal{N}_\bullet, v$ are behaviorally equivalent, then $\mathcal{M} \Vdash \alpha[w]$ iff $\mathcal{N} \Vdash \alpha[v]$ (invariance under monotonic bisimulations can be defined similarly). Furthermore, Proposition 2.16 implies that we can work relative to N while still preserving nice first-order properties such as compactness and the existence of countably saturated models.

Using the translation defined in this section, I can clarify the relationship between non-normal modal logic and first-order logic. For normal modal logic, the seminal *van Benthem characterization theorem* (see Blackburn et al. 2001, Sect. 2.4, for details) shows that normal modal logic is the bisimulation invariant fragment of first-order logic (with respect to relational models). Pauly (1999) generalized this result to monotonic modal logics, showing that, over the class of N^{mon} of $\mathcal{L}_{fo}$-models isomorphic to monotonic neighborhood models, $\alpha(x)$ is equivalent to a translation of a basic modal formula iff $\alpha(x)$ is invariant under monotonic bisimulations (see, also, Hansen 2003). A similar result can be shown for *all* non-normal modal logics using the notion of behavioral equivalence.

Theorem 2.68 (Hansen et al. 2009) *Suppose that* N *is the class of* $\mathcal{L}_{fo}$*-structures isomorphic to the translation of neighborhood models and that* $\alpha(x)$ *is a* $\mathcal{L}_{fo}$*-formula. Then,* $\alpha(x)$ *is equivalent to a translation of a basic modal formula (with respect to* N*) iff* $\alpha(x)$ *is invariant under behavioral equivalence.*

Consult Hansen et al. (2009), Hansen (2003) and Kracht and Wolter (1999) for further model-theoretic results about non-normal modal logic.

Chapter 3
Richer Languages

In the previous chapters, we focused on the basic propositional modal language interpreted on neighborhood structures. One exception was the logic of evidence and belief introduced in Sect. 1.4.4. This logic is interesting, in part, because it includes a non-normal modality (the evidence modality) and two normal modalities (the universal modality and the belief modality). In this final chapter, we will look more systematically at different extensions of the basic modal language interpreted on neighborhood structures.

3.1 Universal Modality and Nominals

A very useful and well-studied extension of the basic modal language includes the universal modality and *nominals*.[1] We have already studied logics with the universal modality. Nominals provide additional expressive power to "name" possible worlds (see Areces and ten Cate 2007 for a discussion). Recall that At is the set of atomic propositions, and let Nom be an additional set of atomic propositions (assume that $\text{At} \cap \text{Nom} = \emptyset$). Elements of Nom are called *nominals*. The key feature of nominals is that they are true at exactly one world—i.e., they "name" possible worlds in a neighborhood model. Let $\mathcal{L}_{NA}$ be the smallest set of formulas generated by the following grammar:

$$p \mid i \mid \neg \varphi \mid (\varphi \wedge \psi) \mid \Box \varphi \mid [A]\varphi$$

where $p \in \text{At}$ and $i \in \text{Nom}$. Let $\mathcal{L}_A$ be the fragment of $\mathcal{L}_{NA}$ without nominals. A **neighborhood model with nominals** is a tuple $\langle W, N, V \rangle$, where $\langle W, N \rangle$ is a

---

[1] See Goranko and Passy (1992) for a discussion of the universal modality and Areces and ten Cate (2007) for a discussion of nominals in modal logic.

neighborhood frame, and $V : \text{At} \cup \text{Nom} \to \wp(W)$ is a valuation function satisfying the property: For all $i \in \text{Nom}$, $|V(i)| = 1$.

To simplify notation, I use $i, j, k, \ldots$ to denote nominals and write $V(i) = w$ when $V(i) = \{w\}$. The interpretation of the Boolean connectives and the modal operator is as usual (see Definition 1.12). I only give the definition of truth for nominals and the modalities. Let $\mathcal{M} = \langle W, N, V \rangle$ be a neighborhood function with nominals. Then,

- $\mathcal{M}, w \models i$ iff $i \in \text{Nom}$ and $V(i) = w$
- $\mathcal{M}, w \models [A]\varphi$ iff for all $v \in W$, $\mathcal{M}, v \models \varphi$
- $\mathcal{M}, w \models \Box\varphi$ iff $[\![\varphi]\!]_\mathcal{M} \in N(w)$

It is well-known that the universal modality cannot be expressed in the basic modal language over relational models. A similar argument shows that the universal modality cannot be expressed in the basic modal language over neighborhood models.

Exercise 3.1 Use the definition of a disjoint union of models (Definition 2.10) to prove that the basic modal language cannot express the universal modality.

There are two natural questions that one can ask when studying extensions of the basic modal language. First, do various properties of the logical system (e.g., completeness, decidability, complexity of the satisfiability problem, etc.) also apply to the extended language? I do not have the space to discuss all aspects of this question, so, instead, I focus on only two issues. In the next section, I show that strong completeness for monotonic modal logics (Theorem 2.41) can be extended to languages with the universal modality. Second, what more can be expressed in the language? Again, I do not have the space to discuss the full expressive power of the modal languages with nominal and universal modality on neighborhood structures. Consult ten Cate et al. (2009) for a comprehensive discussion of expressivity of this language on topological models. In Sect. 3.1.2, I discuss a fascinating result showing that modal languages with nominals can express that a neighborhood frame is augmented.

3.1.1 Non-normal Modal Logic with the Universal Modality

The method for proving completeness, discussed in Sect. 2.3.2, can be adapted to deal with logics that include both normal and non-normal modalities. In this section, I prove completeness for a monotonic modal logic with the universal modality. Since I am restricting attention to monotonic neighborhoods, I will use $\langle\,]$ to denote the modality (cf. the discussion in Sect. 1.2.2). In this case, the language $\mathcal{L}_A$ is the smallest set of formulas generated by the following grammar:

$$p \mid \neg\varphi \mid (\varphi \wedge \psi) \mid \langle\,]\varphi \mid [A]\varphi$$

3.1 Universal Modality and Nominals

where $p \in \mathsf{At}$. The additional Boolean connectives are defined as usual. Furthermore, let $[\ )\varphi$ be defined as $\neg\langle\ ]\neg\varphi$ and $\langle A\rangle\varphi$ be defined as $\neg[A]\neg\varphi$. The interpretation of formulas from $\mathcal{L}_A$ is provided in the previous section. As a reminder, the truth clause neighborhood modality is:

- $\mathcal{M}, w \models \langle\ ]\varphi$ iff there exists $X \in N(w)$ such that $X \subseteq [\![\varphi]\!]_\mathcal{M}$.

Let **EMA** be the logic consisting of the following axiom schemes and rules:

(AK) $[A](\varphi \to \psi) \to ([A]\varphi \to [A]\psi)$
(AT) $[A]\varphi \to \varphi$
(A4) $[A]\varphi \to [A][A]\varphi$
(A5) $\langle A\rangle\varphi \to [A]\langle A\rangle\varphi$
(ANec) From φ infer $[A]\varphi$
($\langle\ ]$RM)   From $\varphi \to \psi$ infer $\langle\ ]\varphi \to \langle\ ]\psi$
($\langle\ ]$Cons) $\neg\langle\ ]\bot$
(AN) $[A]\varphi \to \langle\ ]\varphi$
(Pullout) $\langle\ ](\varphi \wedge [A]\psi) \leftrightarrow (\langle\ ]\varphi \wedge [A]\psi)$

Lemma 3.1 *The axiom* (Pullout) *is valid on any neighborhood model* $\mathcal{M} = \langle W, N, V\rangle$ *in which, for all* $w \in W$, $\emptyset \notin N(w)$.

Proof As the reader is invited to verify, for any formula α of $\mathcal{L}_A$ and any neighborhood model $\mathcal{M} = \langle W, N, V\rangle$, we have that

$$[\![[A]\alpha]\!]_\mathcal{M} = \begin{cases} W & [\![\alpha]\!]_\mathcal{M} = W \\ \emptyset & [\![\alpha]\!]_\mathcal{M} \neq W \end{cases}$$

Let $\mathcal{M} = \langle W, N, V\rangle$ be a neighborhood model such that for all $w \in W$, $\emptyset \notin N(w)$. We must show that $[\![\langle\ ](\varphi \wedge [A]\psi)]\!]_\mathcal{M} = [\![\langle\ ]\varphi \wedge [A]\psi]\!]_\mathcal{M}$. There are two cases:

1. $[\![\psi]\!]_\mathcal{M} = W$. Then, $[\![[A]\psi]\!]_\mathcal{M} = W$. Since $[\![\varphi]\!]_\mathcal{M} \subseteq W = [\![[A]\psi]\!]_\mathcal{M}$ and $[\![\langle\ ]\varphi]\!]_\mathcal{M} \subseteq W = [\![[A]\psi]\!]_\mathcal{M}$, we have that (i) $[\![\varphi]\!]_\mathcal{M} \cap [\![[A]\psi]\!]_\mathcal{M} = [\![\varphi]\!]_\mathcal{M}$ and (ii) $[\![\langle\ ]\varphi]\!]_\mathcal{M} \cap [\![[A]\psi]\!]_\mathcal{M} = [\![\langle\ ]\varphi]\!]_\mathcal{M}$. Then,

$$\begin{aligned}
[\![\langle\ ](\varphi \wedge [A]\psi)]\!]_\mathcal{M} &= \{w \mid \text{there is a } X \in N(w) \text{ such that } X \subseteq [\![\varphi \wedge [A]\psi]\!]_\mathcal{M}\} \\
&= \{w \mid \text{there is a } X \in N(w) \text{ such that } X \subseteq [\![\varphi]\!]_\mathcal{M} \cap [\![[A]\psi]\!]_\mathcal{M}\} \\
&= \{w \mid \text{there is a } X \in N(w) \text{ such that } X \subseteq [\![\varphi]\!]_\mathcal{M}\} \\
&= [\![\langle\ ]\varphi]\!]_\mathcal{M} \\
&= [\![\langle\ ]\varphi]\!]_\mathcal{M} \cap [\![[A]\psi]\!]_\mathcal{M} \\
&= [\![\langle\ ]\varphi \wedge [A]\psi]\!]_\mathcal{M}
\end{aligned}$$

2. $[\![\psi]\!]_\mathcal{M} \neq W$. Then, $[\![[A]\psi]\!]_\mathcal{M} = \emptyset$. Thus, $[\![\varphi \wedge [A]\psi]\!]_\mathcal{M} = [\![\langle\ ]\varphi \wedge [A]\psi]\!]_\mathcal{M} = \emptyset$. Since, $\emptyset \notin N(w)$ for any w, we have that $[\![\langle\ ](\varphi \wedge [A]\psi)]\!]_\mathcal{M} = \emptyset = [\![\langle\ ]\varphi \wedge [A]\psi]\!]_\mathcal{M}$. $\square$

Exercise 3.2 Prove that (AN) is valid on any neighborhood model $\mathcal{M} = \langle W, N, V\rangle$ in which for all $w \in W$, $W \in N(w)$.

The completeness proof combines ideas from the previous section and the standard method for proving completeness with respect to relational frames. Let $\mathsf{F}_{EMA} = \{\mathcal{F} \mid \mathcal{F} = \langle W, N \rangle$ with for all $w \in W, \emptyset \notin N(w)$ and $W \in N(w)\}$. Recall that $M_{\mathbf{EMA}}$ is the set of maximally **EMA**-consistent sets. Finally, suppose that R^A is a relation on $M_{\mathbf{EMA}}$ defined as follows:

$$\Gamma \; R^A \; \Delta \text{ iff } \Gamma^A = \{\varphi \mid [A]\varphi \in \Gamma\} \subseteq \Delta.$$

The following lemma is a consequence of the fact that **EMA** contains the axiom schemes (**AT**), (**A4**) and (**A5**) (see Blackburn et al. 2001, for details).

Lemma 3.2 *The relation R^A is an equivalence relation.*

The proof of this standard fact about modal logic is left to the reader. Let Γ be a maximally consistent set. For any maximally consistent set $\Gamma \in M_{\mathbf{EMA}}$, construct a canonical model $\mathcal{M}^\Gamma = \langle W^\Gamma, N^\Gamma, V^\Gamma \rangle$ as follows:

- $W^\Gamma = R^A(\Gamma) = \{\Delta \mid \Gamma \; R^A \; \Delta\}$;
- For each $\Delta \in W^\Gamma$, $N^\Gamma(\Delta) = \{|\varphi|_{\mathbf{EMA}} \cap W^\Gamma \mid \langle \,]\varphi \in \Delta\}$; and
- For all $p \in \mathsf{At}$, $V^\Gamma(p) = \{\Delta \in W^\Gamma \mid p \in \Delta\} = |p|_{\mathbf{EMA}} \cap W^\Gamma$.

Before proving a Truth Lemma for this canonical model, we need a preliminary result about the logic **EMA**.

Lemma 3.3 *Suppose that Γ is a set of formulas. If $\Gamma, \varphi \vdash_{\mathbf{EMA}} \psi$, then $[A]\Gamma, \langle \,]\varphi \vdash_{\mathbf{EMA}} \langle \,]\psi$, where $[A]\Gamma = \{[A]\varphi \mid \varphi \in \Gamma\}$.*

Proof Without loss of generality, we can replace Γ with a single formula γ (why?). Suppose that $\gamma, \varphi \vdash_{\mathbf{EMA}} \psi$. Then, by the definition of a deduction, we have that $\vdash_{\mathbf{EMA}} (\gamma \wedge \varphi) \to \psi$. Since $[A]\gamma \to \gamma$ is an instance of (**AT**), we also have $\vdash_{\mathbf{EMA}} ([A]\gamma \wedge \varphi) \to \psi$. Thus, by $(\langle \,]\mathsf{RM})$, $\vdash_{\mathbf{EMA}} \langle \,]([A]\gamma \wedge \varphi) \to \langle \,]\psi$. Using the axiom (**Pullout**), we have that $\vdash_{\mathbf{EMA}} ([A]\gamma \wedge \langle \,]\varphi) \to \langle \,]\psi$. By the definition of a deduction, this means that $[A]\gamma, \langle \,]\varphi \vdash_{\mathbf{EMA}} \langle \,]\psi$, as desired. □

Lemma 3.4 (Truth Lemma) *Suppose that Γ is a maximally consistent set. For any formula $\varphi \in \mathcal{L}^{MA}$,*

$$[\![\varphi]\!]_{\mathcal{M}^\Gamma} = |\varphi|_{\mathbf{EMA}} \cap R^A(\Gamma).$$

Proof Suppose that Γ is a maximally consistent set and $\mathcal{M}^\Gamma$ is the canonical model for Γ. The proof is by induction on the structure of $\varphi \in \mathcal{L}^{MA}$. The base case is a direct consequence of the definition of the canonical valuation:

$$[\![p]\!]_{\mathcal{M}^\Gamma} = V^\Gamma(p) = |p|_{\mathbf{EMA}} \cap R^A(\Gamma).$$

The proof for the Boolean connectives are straightforward. I give the details only for the modal operators:

3.1 Universal Modality and Nominals

- $[\![\langle\,]\varphi]\!]_{\mathcal{M}^\Gamma} = |\langle\,]\varphi|_{\mathbf{EMA}} \cap R^A(\Gamma)$: Suppose that $\Delta \in |\langle\,]\varphi|_{\mathbf{EMA}} \cap R^A(\Gamma)$. Then, $\langle\,]\varphi \in \Delta$ and $\Delta \in R^A(\Gamma)$. By construction, $|\varphi| \cap R^A(\Gamma) \in N^\Gamma(\Delta)$. Hence, by the induction hypothesis, $[\![\varphi]\!]_{\mathcal{M}^\Gamma} = |\varphi| \cap R^A(\Gamma) \in N^\Gamma(\Delta)$. Thus, $\Delta \in [\![\langle\,]\varphi]\!]_{\mathcal{M}^\Gamma}$.
  Suppose that $\Delta \notin |\langle\,]\varphi|_{\mathbf{EMA}} \cap R^A(\Gamma)$. If $\Delta \notin R^A(\Gamma)$, then obviously $\Delta \notin [\![\langle\,]\varphi]\!]_{\mathcal{M}^\Gamma}$. So, assume for the remainder of the proof that $\Delta \in R^A(\Gamma)$. Then, $\langle\,]\varphi \notin \Delta$. We must show that $\mathcal{M}, \Delta \not\models \langle\,]\varphi$. That is, we must show that for all $X \in N^\Gamma(\Delta)$, $X \not\subseteq [\![\varphi]\!]_{\mathcal{M}^\Gamma}$. This is a consequence of the following claim:

  Claim. For each ψ with $\langle\,]\psi \in \Delta$, there is a maximally consistent set Δ' such that $\Delta' \in |\psi|_{\mathbf{EMA}} \cap R^A(\Gamma)$, but $\varphi \notin \Delta'$.

  Proof of claim. Let $\Delta'_0 = \Gamma^A \cup \{\neg\varphi\} \cup \{\psi\}$. First of all, since $\langle\,]\varphi \notin \Delta$, using the axiom scheme (**AN**), we have $A\varphi \notin \Delta$. Thus, since $\Delta \in R^A(\Gamma)$, $A\varphi \notin \Gamma$ (why?). Therefore, $\varphi \notin \Gamma^A$. We now show that Δ'_0 is consistent. Suppose not. Then, $\Gamma^A \cup \{\neg\varphi\} \cup \{\psi\} \vdash_{\mathbf{EMA}} \bot$. Using standard propositional reasoning, $\Gamma^A \cup \{\psi\} \vdash_{\mathbf{EMA}} \varphi$. By Lemma 3.3, $[A]\Gamma^A \cup \{\langle\,]\psi\} \vdash_{\mathbf{EMA}} \langle\,]\varphi$. Since $[A]\Gamma^A \subseteq \Gamma^A \subseteq \Delta$ and $\langle\,]\psi \in \Delta$, we have $\langle\,]\varphi \in \Delta$, which is a contradiction. Thus, Δ'_0 is consistent. By Lindenbaum's Lemma,[2] there is a maximally consistent set Δ' such that $\Delta'_0 \subseteq \Delta'$. Therefore, $\Delta' \in |\psi|_{\mathbf{EMA}} \cap R^A(\Gamma)$ but $\varphi \notin \Delta'$. This completes the proof of the claim.

- $[\![[A]\varphi]\!]_{\mathcal{M}^\Gamma} = |[A]\varphi|_{\mathbf{EMA}} \cap R^A(\Gamma)$: Suppose that $\Delta \in |[A]\varphi|_{\mathbf{EMA}} \cap R^A(\Gamma)$. Then, in particular, $[A]\varphi \in \Delta$. Let $\Delta' \in R^A(\Gamma)$. Then, since $\Delta \in R^A(\Gamma)$, we have $\Delta\, R^A \Delta'$. Hence, since $[A]\varphi \in \Delta$ and $\Delta^A \subseteq \Delta'$, we have $\varphi \in \Delta'$. Thus, $R^A(\Gamma) \subseteq |\varphi|_{\mathbf{EMA}}$. Then, by the induction hypothesis, $[\![\varphi]\!]_{\mathcal{M}^\Gamma} = |\varphi|_{\mathbf{AMA}} \cap R^A(\Gamma) = R^A(\Gamma)$. Thus, $\Delta \in [\![[A]\varphi]\!]_{\mathcal{M}^\Gamma}$.
  Suppose that $\Delta \notin |[A]\varphi|_{\mathbf{EMA}} \cap R^A(\Gamma)$. Again, we can assume that $\Delta \in R^A(\Gamma)$. Then, $[A]\varphi \notin \Delta$. We must show that there is a $\Delta' \in R^A(\Gamma)$ such that $\varphi \notin \Delta'$. Let $\Delta'_0 = \Gamma^A \cup \{\neg\varphi\}$. We claim that Δ'_0 is consistent. Suppose not. Then, $\Gamma^A \cup \{\neg\varphi\} \vdash_{\mathbf{EMA}} \bot$. Using standard propositional reasoning, there are formulas $\alpha_1, \ldots, \alpha_n \in \Gamma^A$ such that $\vdash_{\mathbf{EMA}} (\alpha_1 \wedge \cdots \wedge \alpha_n) \to \varphi$. Using (**ANec**) and (**AK**), standard modal reasoning gives us $\vdash_{\mathbf{EMA}} ([A]\alpha_1 \wedge \cdots \wedge [A]\alpha_n) \to [A]\varphi$. Hence, $([A]\alpha_1 \wedge \cdots \wedge [A]\alpha_n) \to [A]\varphi \in \Delta$. Since, for each $i = 1, \ldots, n$, $[A]\alpha_i \in \Gamma$, we conclude that $[A]\alpha_i \in \Delta$ for each $i = 1, \ldots, n$. Thus, $[A]\varphi \in \Delta$, which is a contradiction. Therefore, Δ'_0 is consistent. By Lindenbaum's Lemma, there is a maximally **EMA**-consistent set Δ' such that $\Delta'_0 \subseteq \Delta'$. Then, $\Delta' \notin |\varphi|_{\mathbf{EMA}} \cap R^A(\Gamma) = [\![\varphi]\!]_{\mathcal{M}^\Gamma}$. Thus, $\Delta \notin [\![[A]\varphi]\!]_{\mathcal{M}^\Gamma}$. □

Strong completeness now follows by the standard argument.

Theorem 3.1 *The logic* **EMA** *is sound and strongly complete with respect to class of consistent, monotonic neighborhood frames.*

---

[2] It should also be verified that Lindenbaum's Lemma holds for **EMA**. I leave the verification of this to the reader.

3.1.2 Characterizing Augmented Frames

Recall that a frame $\mathcal{F} = \langle W, N \rangle$ is augmented provided that for each $w \in W$, $N(w)$ is non-empty, $N(w)$ is a filter and $\bigcap N(w) \in N(w)$. This last property (that the neighborhoods contain its core) is not expressible in the basic modal language.

Exercise 3.3 Use the definition of a monotonic bisimulation to prove that, over the class of frames that are filters, the basic modal language cannot express that a neighborhood frame is augmented.

Interestingly, ten Cate and Litak (2007) showed that within the class of neighborhood frames that are filters, there is a sense in which the augmented frames can be defined. They use a somewhat non-standard approach and define the class of augmented frames using a *rule*. To do this, we need to understand the notion of *admitting a rule*. A rule of inference is *admitted* in a logic **L** if adding it to the logic does not add any new theorems. There is an important distinction between *deriving* a rule and *admitting* a rule. A rule is derivable if the consequent can be derived in the logic from the premises. To illustrate, consider the rule: from $\Box \varphi$ infer φ. This rule obviously cannot be derived in **K**, but it is admitted, since it does not add any new theorems to **K**. There is a semantic characterization of admitting a rule that is needed to state the main theorem of this section.

Definition 3.2 (*Admitting a Rule*) Suppose that $\mathcal{M} = \langle W, N, V \rangle$ is a neighborhood model with a state satisfying a formula φ. A model $\mathcal{M}'$ φ-*extends* the model $\mathcal{M}$ provided that $\mathcal{M}' = \langle W, N, V' \rangle$, where V' is the same as V for all atomic propositions and nominals that occur in φ. A class of frames **F** **admits** a rule provided that every model based on a frame from **F** that falsifies the consequent γ can be γ-extended to a model that falsifies the premises.

It turns out that a rule, familiar in the literature on hybrid logic (Areces and ten Cate 2007), can be used to characterize augmented frames:

$$(\mathbf{BG}) \quad \frac{\langle A \rangle (i \wedge \Diamond j) \to \langle A \rangle (j \wedge \varphi)}{\langle A \rangle (i \wedge \Box \varphi)}$$

for $i \neq j$ and j not occurring in φ.

Proposition 3.1 (ten Cate and Litak 2007) *Let* $\mathcal{F} = \langle W, N \rangle$ *be a neighborhood frame such that for all* $w \in W$, $N(w)$ *is a non-trivial filter. Then,* $\mathcal{F}$ *contains its core (i.e., for all* $w \in W$, $\bigcap N(w) \in N(w)$*) if, and only if,* $\mathcal{F}$ *admits the* (**BG**) *rule.*

Proof Suppose that $\mathcal{F} = \langle W, N \rangle$ is an augmented frame. That is, for all $w \in W$, $N(w)$ is a filter and $\bigcap N(w) \in N(w)$. Suppose that $\mathcal{M} = \langle W, N, V \rangle$ is a model that falsifies $\langle A \rangle (i \wedge \Box \varphi)$. Then, there is a $w \in W$ such that $V(i) = w$ and $\mathcal{M}, w \not\models \Box \varphi$. Thus, for each $X \in N(w)$, $X \neq [\![\varphi]\!]_{\mathcal{M}}$. Since $N(w)$ is a filter and $\bigcap N(w)$ is the smallest element of $N(w)$ (according to the subset relation), this means that there

3.1 Universal Modality and Nominals

must be a $v \in \bigcap N(w)$ such that $v \notin [\![\varphi]\!]_{\mathcal{M}}$. Let $V' : \text{At} \cup \text{Nom} \to \wp(W)$ be a valuation that is exactly like V except $V'(j) = v$, where j is a nominal that does not occur in φ and $j \neq i$. Note that since $v \notin \bigcap N(w)$ and $N(w)$ is closed under supersets, we have $W - \{v\} \in N(w)$. Now, $\mathcal{M}' = \langle W, N, V' \rangle$ is a model that $\langle A \rangle (i \wedge \Box \varphi)$-extends $\mathcal{M}$ falsifying the antecedent:

1. $\mathcal{M}', w \models \langle A \rangle (i \wedge \Diamond j)$: This follows from the fact that $W - \{v\} \in N(w)$ and $[\![\Diamond j]\!]_{\mathcal{M}'} = \{x \mid W - \{v\} \notin N(x)\}$; but
2. $\mathcal{M}', w \not\models \langle A \rangle (j \wedge \varphi)$: By construction $V'(j) = v \notin [\![\varphi]\!]_{\mathcal{M}} = [\![\varphi]\!]_{\mathcal{M}'}$ (the latter follows from the fact that V and V' agree on all formulas not involving j).

Suppose that $\mathcal{F} = \langle W, N \rangle$ is not augmented. Then, there is a $w \in W$ such that $\bigcap N(w) \notin N(w)$. Since $N(w)$ is a filter, this means that for each $X \in N(w)$ there is a neighborhood $X' \in N(w)$ such that $X \not\subseteq X'$. Thus, for each $X \in N(w)$, we have

$$g(X) = \{y \mid y \in X, \text{ and there is an } X' \in N(w) \text{ such that } y \notin X'\} \neq \emptyset.$$

Thus, by the *Axiom of Choice*,[3] for each $X \in N(w)$ we can choose an element $y_X \in g(X)$. Now, consider the set $\{y_X \mid X \in N(w)\}$. There are two key observations about this set:

1. For each $X \in N(w)$, $W - \{y_X\} \in N(w)$: since $y_X \notin X'$ for some $X' \in N(w)$ and $N(w)$ is closed under supersets, we have $X' \subseteq W - \{y_X\} \in N(w)$. Thus, if $\mathcal{M}$ is a model with a valuation V, where $V(j) = y_X$ for some $j \in \text{Nom}$, then $\mathcal{M}, w \models \Box \neg j$; and so, $\mathcal{M}, w \not\models \Diamond j$.
2. $W - \{y_X \mid X \in N(w)\} \notin N(w)$: Since for each $X \in N(w)$, $y_X \in X$, there is no $X' \in N(w)$ such that $X' = W - \{y_X \mid X \in N(w)\}$. Thus, if $\mathcal{M}$ is a model based on the frame $\mathcal{F}$ in which $[\![\varphi]\!]_{\mathcal{M}} = W - \{y_X \mid X \in N(w)\}$, then $\mathcal{M}, w \not\models \Box \varphi$.

Let $\mathcal{M} = \langle W, N, V \rangle$ be a model based on the frame $\mathcal{F}$ with $V(i) = w$ and $V(p) = W - \{y_X \mid X \in N(w)\}$. Then, $\mathcal{M}$ is a falsifying model for $\langle A \rangle (i \wedge \Box p)$. By item 2, above, $\mathcal{M}, w \not\models \Box p$; and so, $\mathcal{M}, w \not\models i \wedge \Box p$ (since $V(i) = w$, we have $[\![\langle A \rangle (i \wedge \Box \varphi)]\!]_{\mathcal{M}} = \emptyset$).

Let $\mathcal{M}' = \langle W, N, V' \rangle$ be any model in which V' is a valuation on $\mathcal{F}$ such that $V'(i) = V(i) = w$ and $V'(p) = V(p) = W - \{y_X \mid X \in N(w)\}$. Then, $\mathcal{M}'$ is a model that $\langle A \rangle (i \wedge \Box \varphi)$-extends $\mathcal{M}$. The claim is that $\mathcal{M}'$ is not a falsifying model for $\langle A \rangle (i \wedge \Diamond j) \to \langle A \rangle (j \wedge p)$, where $j \neq i$. There are two cases:

1. $V'(j) = y_X$ for some $X \in N(w)$. Then, as noted in item 1, above, $\mathcal{M}, w \not\models \Diamond j$. Thus, $[\![\langle A \rangle (i \wedge \Diamond j)]\!]_{\mathcal{M}'} = \emptyset$. Hence, $[\![\langle A \rangle (i \wedge \Diamond j) \to \langle A \rangle (j \wedge p)]\!]_{\mathcal{M}'} = W$.
2. $V'(j) \neq y_X$ for any $X \in N(w)$. Then, since $V'(j) \in [\![p]\!]_{\mathcal{M}'} = W - \{y_X \mid X \in N(w)\}$, we have $\mathcal{M}, w \models \langle A \rangle (j \wedge p)$. Hence, $[\![\langle A \rangle (i \wedge \Diamond j) \to \langle A \rangle (j \wedge p)]\!]_{\mathcal{M}'} = W$.

In either case, $\mathcal{M}'$ is not a falsifying model for $\langle A \rangle (i \wedge \Diamond j) \to \langle A \rangle (j \wedge p)$. □

---

[3]The Axiom of Choice says that for any collection of non-empty sets indexed by any set I, $(X_i)_{i \in I}$, there is a sequence $(x_i)_{i \in I}$ of elements such that for all $i \in I$, $x_i \in X_i$.

3.2 First-Order Neighborhood Structures

3.2.1 Syntax and Semantics

The first-order modal language includes features from propositional modal logic and first-order logic. For simplicity, the language in this section will not include constants, terms or equality. Suppose that $\mathcal{V}$ is a countable set of variables. For each natural number $n \geq 0$, there is a (countable) set of predicate symbols of arity n. Predicate symbols will be denoted by capital letters $F^{(n)}, G^{(n)}, \ldots$. To simplify the notation, I will write F instead of $F^{(n)}$ when the arity of F is understood. The set of first-order modal formulas, denoted $\mathcal{L}_1$, is the smallest set of formulas generated by the grammar:

$$F(x_1, \ldots, x_n) \mid \neg\varphi \mid (\varphi \wedge \psi) \mid \Box\varphi \mid \forall x \varphi$$

where F is an n-ary predicate symbol, $x \in \mathcal{V}$, and for $i = 1, \ldots, n$, $x_i \in \mathcal{V}$. The other Boolean connectives, the diamond modal operator ($\Diamond$) and the existential quantifier ($\exists$) are defined in the standard way. For instance, $\exists x \varphi$ is defined as $\neg \forall x \neg \varphi$. The usual rules about free variables apply. For a formula $\varphi \in \mathcal{L}_1$, let $Fr(\varphi)$ denote the set of free variables in φ. I write $\varphi(x)$ when x (possibly) occurs free in φ. The formula $\varphi[y/x]$ is φ, in which the free variable x is replaced with a variable y. The variable y is **substitutable for** x **in** φ provided that y is free in $\varphi[y/x]$.[4]

Adding first-order structures to neighborhood frames and models is straightforward.

Definition 3.3 (*Constant Domain Neighborhood Frames*) A **constant-domain neighborhood frame** is a tuple $\langle W, N, D \rangle$, where W is a non-empty set of possible worlds; $N : W \to \wp(\wp(W))$ is a neighborhood function; and D is a set (called the **domain**).

Definition 3.4 (*Constant Domain Neighborhood Models*) Let $\mathcal{F} = \langle W, N, D \rangle$ be a constant-domain neighborhood frame. A **constant-domain neighborhood model** based on $\mathcal{F}$ is a tuple $\langle W, N, D, I \rangle$, where I is a first-order interpretation function: for all n-ary predicate symbols F, $I(F, w) \subseteq D^n$.

Remark 3.5 (*Variable Domain Models*) It is also common to consider **variable domain models** in which different domains are assigned to different possible worlds. There is no inherent difficulty in defining variable domain neighborhood frames/models. It would be interesting to extend the results discussed in this section to this more general setting (cf. Awodey and Kishida 2008; Kishida 2011; Calardo 2013).

Since formulas of $\mathcal{L}_1$ may include free variables, truth is defined at a state and an assignment. An **assignment** is a function assigning elements of the domain to

---

[4] The usual restrictions apply when substituting y for x in φ so that either bound variables are renamed or y is assumed to be substitutable for x in φ. See Enderton (2001) for a discussion.

3.2 First-Order Neighborhood Structures

variables: $\sigma : \mathcal{V} \to D$. The following definition is needed to define truth of quantified formulas.

Definition 3.6 (*x-variant*) An *x*-**variant** of an assignment σ is an assignment σ' such that for all $y \in \mathcal{V}$, if $y \neq x$, then $\sigma(y) = \sigma'(y)$. Write $\sigma \sim_x \sigma'$ when σ' is an *x*-variant of σ.

Definition 3.7 (*Truth*) Suppose that $\mathcal{M} = \langle W, N, D, I \rangle$ is a constant-domain neighborhood model and σ is an assignment. For $\varphi \in \mathcal{L}_1$, truth at state $w \in W$ with respect to σ, denoted $\mathcal{M}, w \models_\sigma \varphi$, is defined by induction on the structure of φ:

1. $\mathcal{M}, w \models_\sigma F(x_1, \ldots, x_n)$ iff $\langle \sigma(x_1), \ldots, \sigma(x_n) \rangle \in I(F, w)$, where F is an n-place predicate symbol.
2. $\mathcal{M}, w \models_\sigma \neg \varphi$ iff $\mathcal{M}, w \not\models_\sigma \varphi$
3. $\mathcal{M}, w \models_\sigma \varphi \wedge \psi$ iff $\mathcal{M}, w \models_\sigma \varphi$ and $\mathcal{M}, w \models_\sigma \psi$
4. $\mathcal{M}, w \models_\sigma \Box \varphi$ iff $[\![\varphi]\!]_{\mathcal{M}, \sigma} \in N(w)$
5. $\mathcal{M}, w \models_\sigma \forall x \varphi(x)$ iff for each σ', if $\sigma \sim_x \sigma'$, then $\mathcal{M}, w \models_{\sigma'} \varphi(x)$,

where $[\![\varphi]\!]_{\mathcal{M}, \sigma} = \{w \mid \mathcal{M}, w \models_\sigma \varphi\} \subseteq W$.

Example 3.8 (*Detailed Example of a First-Order Neighborhood Model*) Suppose that F is a unary predicate symbol, $\mathcal{V} = \{x, y\}$, and $\langle W, N, D, I \rangle$ is a first-order constant-domain neighborhood model, where

- $W = \{w, v, u\}$;
- $N(w) = \{\{w, v\}, \{v, u\}\}$, $N(v) = \{\{v\}\}$, $N(u) = \{\{w, v\}, \{v\}\}$;
- $D = \{a, b\}$; and
- $I(F, w) = \{a\}$, $I(F, v) = \{a, b\}$, and $I(F, u) = \emptyset$.

There are four possible assignments:

- $\sigma_1 : \mathcal{V} \to D$, where $\sigma_1(x) = a$, $\sigma_1(y) = b$;
- $\sigma_2 : \mathcal{V} \to D$, where $\sigma_2(x) = b$, $\sigma_2(y) = a$;
- $\sigma_3 : \mathcal{V} \to D$, where $\sigma_3(x) = \sigma_3(y) = a$; and
- $\sigma_4 : \mathcal{V} \to D$, where $\sigma_4(x) = \sigma_4(y) = b$

As the reader is invited to verify, we have the following truth sets associated with the atomic formula $F(x)$:

- $[\![F(x)]\!]_{\mathcal{M}, \sigma_1} = \{w, v\}$;
- $[\![F(x)]\!]_{\mathcal{M}, \sigma_2} = \{v\}$;
- $[\![F(x)]\!]_{\mathcal{M}, \sigma_3} = \{w, v\}$; and
- $[\![F(x)]\!]_{\mathcal{M}, \sigma_4} = \{v\}$.

In general, every formula $\varphi \in \mathcal{L}_1$ is associated with a function

$$[\![\varphi]\!] : D^\mathcal{V} \to \wp(W)$$

assigning sets of states to each assignment (note that $D^\mathcal{V}$ is the set of all functions from $\mathcal{V}$ to D). Notice that $F(x)$ and $F(y)$ are different formulas (since they have different free variables), thus, are associated with different functions (although the ranges of these functions are the same). Finally, the reader is invited to verify the following:

- $[\![\Box F(x)]\!]_{\mathcal{M},\sigma_1} = [\![\Box F(x)]\!]_{\mathcal{M},\sigma_3} = \{w, u\}$ and $[\![\Box F(x)]\!]_{\mathcal{M},\sigma_2} = [\![\Box F(x)]\!]_{\mathcal{M},\sigma_4} = \{v\}$
- $[\![\Box \forall x F(x)]\!]_{\mathcal{M},\sigma_1} = [\![\Box \forall x F(x)]\!]_{\mathcal{M},\sigma_2} = [\![\Box \forall x F(x)]\!]_{\mathcal{M},\sigma_3} = [\![\Box \forall x F(x)]\!]_{\mathcal{M},\sigma_4} = \{v, u\}$
- $[\![\forall x \Box F(x)]\!]_{\mathcal{M},\sigma_1} = [\![\forall x \Box F(x)]\!]_{\mathcal{M},\sigma_2} = [\![\forall x \Box F(x)]\!]_{\mathcal{M},\sigma_3} = [\![\forall x \Box F(x)]\!]_{\mathcal{M},\sigma_4} = \{u\}$

Exercise 3.4 For each of the following formulas, find a first-order constant-domain neighborhood model that makes the formula true and one that makes the formula false.

1. $\forall x \Box P(x) \to \Box \forall x P(x)$
2. $\Box \forall x P(x) \to \forall x \Box P(x)$
3. $\exists x \Box P(x) \to \Box \exists x P(x)$
4. $\Box \exists x P(x) \to \exists x \Box P(x)$

3.2.2 The Barcan and Converse Barcan Schema

Discussion of first-order modal logic is most interesting when it is focused on the interaction between modal operators and quantifiers. Two of the most widely discussed axiom schemes allowing interaction between the modal operators and the quantifiers are the Barcan and converse Barcan formulas.

Definition 3.9 (*Barcan/Converse Barcan Schemas*) The Barcan schema is:

$$\text{(BF)} \qquad \forall x \Box \varphi(x) \to \Box \forall x \varphi(x).$$

Instances of (BF) are called **Barcan formulas**. The **converse Barcan schema** is:

$$\text{(CBF)} \qquad \Box \forall x \varphi(x) \to \forall x \Box \varphi(x).$$

Any instance of (CBF) called a converse Barcan formula.

I start by surveying how the (BF) and (CBF) formulas behave on relational first order structures. Recall that a relational frame is a tuple $\langle W, R \rangle$ where W is a nonempty set and $R \subseteq W \times W$ is a relation (cf. Appendix A). A constant-domain first-order relational model based on a relational frame $\mathcal{F} = \langle W, R \rangle$ is a tuple $\langle W, R, D, I \rangle$ where D is a set and I is a first-order classical interpretation assigning n-ary relations to n-ary predicate symbols.

Given a first-order constant-domain relational model with a state w and assignment σ, truth of formulas from $\mathcal{L}_1$ is defined as above (Definition 3.7) except for the modal clause:

3.2 First-Order Neighborhood Structures

$\mathcal{M}, w \models_\sigma \Box\varphi$ iff for each $w' \in W$, if wRw' then $\mathcal{M}, w' \models_\sigma \varphi$.

The following observation is well-known:

Observation 3.10 *The Barcan and converse Barcan formulas are valid on all first-order relational models with constant domains.*

Exercise 3.5 Prove Observation 3.10. This is a well-known observation (see Hughes and Cresswell 1996, p. 245, and Fitting and Mendelsohn 1999, Sect. 4.9).

Returning to first-order neighborhood frames, Arló-Costa (2002) showed that the Barcan and the Converse Barcan formulas correspond to interesting properties of the neighborhood function. Before reporting these results, we need some definitions. Recall that a neighborhood frame is **non-trivial** provided that for all states w, $N(w) \neq \emptyset$. A first-order neighborhood frame $\mathcal{F} = \langle W, N, D \rangle$ has a **non-trivial domain** when $D \neq \emptyset$.

Exercise 3.6 Prove the following:

- If $D = \emptyset$, then the converse Barcan formula is valid. However, the Barcan formula is not valid (unless $W \in N(w)$ for all w).
- If $|D| = 1$, then $\forall x \Box \varphi(x) \leftrightarrow \Box \forall x \varphi(x)$ is valid.
- If $N(w) = \emptyset$ for all states w, then $\forall x \Box \varphi(x) \leftrightarrow \Box \forall \varphi(x)$ is valid.

To keep the notation at a minimum, say that a first-order neighborhood frame is **trivial** provided that either $N(w) = \emptyset$ for all states w or $|D| \leq 1$.

Proposition 3.2 (Arló-Costa 2002) *Suppose that $\mathcal{F}$ is a constant-domain neighborhood frame. The converse Barcan formula is valid on $\mathcal{F}$ iff either $\mathcal{F}$ is trivial or $\mathcal{F}$ is supplemented.*

Proof Suppose that $\mathcal{F}$ is a constant-domain first-order neighborhood model. If $\mathcal{F}$ is trivial, then, as noted in the above exercise, the converse Barcan formula is valid. Suppose that $\mathcal{F} = \langle W, N, D \rangle$ is a non-trivial, monotonic first-order neighborhood frame and that $\mathcal{M} = \langle W, N, D, I \rangle$ is an arbitrary model based on $\mathcal{F}$. Let $w \in W$ and $\sigma : \mathcal{V} \to D$. We will show that $\mathcal{M}, w \models \Box \forall x \varphi(x) \to \forall x \Box \varphi(x)$. Suppose that $\mathcal{M}, w \models_\sigma \Box \forall x \varphi(x)$. Then, $[\![\forall x \varphi(x)]\!]_{\mathcal{M},\sigma} \in N(w)$. Now, for each σ', if $\sigma \sim_x \sigma'$, then $[\![\forall x \varphi(x)]\!]_{\mathcal{M},\sigma} \subseteq [\![\varphi(x)]\!]_{\mathcal{M},\sigma'}$. Therefore, since $N(w)$ is closed under supersets, for each σ' such that $\sigma \sim_x \sigma'$, we have $[\![\varphi(x)]\!]_{\mathcal{M},\sigma'} \in N(w)$. But this implies that $\mathcal{M}, w \models'_\sigma \Box \varphi(x)$ for all σ' such that $\sigma \sim_x \sigma'$. Hence, $\mathcal{M}, w \models_\sigma \forall x \Box \varphi(x)$. This proves the right-to-left implication.

For the left-to-right implication, we must show that if $\mathcal{F}$ is a non-trivial frame that is not monotonic, then the converse Barcan formula is not valid. Suppose that $\mathcal{F} = \langle W, N, D \rangle$ is not closed under supersets. Then, there is some state w and sets X and Y such that $X \in N(w)$, $X \subseteq Y$, but $Y \notin N(w)$. Let F be a unary predicate symbol. We will construct a model $\mathcal{M}$ based on $\mathcal{F}$, where $\mathcal{M}, w \models_\sigma \Box \forall x F(x)$, but $\mathcal{M}, w \not\models_\sigma \forall x \Box F(x)$.

The interpretation of F is defined as follows. Choose an element $a \in D$ (this is possible since $\mathcal{F}$ is non-trivial). Then, for all $v \in W$,

$$I(F, v) = \begin{cases} D & \text{iff } v \in X \\ \{a\} & \text{iff } v \in Y - X \\ \emptyset & \text{iff } v \in W - X \end{cases}$$

Let σ be any assignment. Since there is at least one element $b \in D$ such that $b \neq a$, we have $[\![\forall x F(x)]\!]_{\mathcal{M},\sigma} = X$. Hence, $\mathcal{M}, w \models_\sigma \Box \forall x F(x)$. However, let $\sigma' : \mathcal{V} \to D$ be an assignment that is the same as σ, except $\sigma'(x) = a$. Then, $\sigma \sim_x \sigma'$ and $[\![F(x)]\!]_{\mathcal{M},\sigma'} = Y \notin N(w)$. Hence, $\mathcal{M}, w \not\models_{\sigma'} \Box F(x)$; and so, $\mathcal{M}, w \not\models_\sigma \forall x \Box F(x)$. □

The Barcan formula also corresponds to interesting properties of the neighborhood function. We first need some notation. Let κ be a cardinal. Recall the definitions from Sect. 1.1. A neighborhood frame is closed under less than or equal to κ intersections if, for each state w and each collection of sets $\{X_i \mid i \in I\}$ where $|I| \leq \kappa$, $\bigcap_{i \in I} X_i \in N(w)$. Also, a neighborhood frame is consistent provided that for all states w, $\emptyset \notin N(w)$. One more definition is needed before we can characterize the frames that validate the Barcan formula.

Definition 3.11 (*Richness*) A consistent first-order neighborhood frame $\langle W, N, D \rangle$ is **rich** provided for all $w \in W$, there are at least as many elements in D as there are in $N(w)$. That is, there is a 1-1 function from $N(w)$ to D.

Proposition 3.3 (Arló-Costa 2002; Arló-Costa and Pacuit 2006) *Suppose that $\mathcal{F}$ is a consistent constant-domain neighborhood frame. Then, the Barcan formula is valid on $\mathcal{F}$ iff either (1) $\mathcal{F}$ is trivial; or (2) if D is a non-empty finite set, then $\mathcal{F}$ is closed under finite intersections, and if D is infinite of size κ (where κ is a cardinal), then $\mathcal{F}$ is rich and closed under $\leq \kappa$ intersections.*

Proof Suppose that $\mathcal{F}$ is a consistent constant-domain first-order neighborhood model. If $\mathcal{F}$ is trivial, then, as noted in the above exercise, the Barcan formula is valid on $\mathcal{F}$. Suppose that $\mathcal{M} = \langle W, N, D, I \rangle$ is a model based on $\mathcal{F}$. Let $w \in W$ and σ be an assignment σ. We must show that $\mathcal{M}, w \models_\sigma \forall x \Box \varphi(x) \to \Box \forall x \varphi(x)$. Suppose that $\mathcal{M}, w \models_\sigma \forall x \Box \varphi(x)$. If D is finite, then $\{[\![\varphi(x)]\!]_{\mathcal{M},\sigma'} \mid \sigma' \sim_x \sigma\}$ is finite. Now, since $[\![\varphi(x)]\!]_{\mathcal{M},\sigma'} \in N(w)$ for each σ' such that $\sigma \sim_x \sigma'$ and N is closed under finite intersections, we have that $\bigcap \{[\![\varphi(x)]\!]_{\mathcal{M},\sigma'} \mid \sigma' \sim_x \sigma\} \in N(w)$. Therefore, $\mathcal{M}, w \models_\sigma \Box \forall x \varphi(x)$. The proof is similar when D is infinite. This shows that the conditions in the proposition imply that the Barcan formula is valid.

For the converse, suppose that $\mathcal{F} = \langle W, N, D \rangle$ is a consistent, non-trivial, constant-domain first-order neighborhood frame such that $|D| \geq 2$. Suppose, now, that D is finite, but that $\mathcal{F}$ is not closed under finite intersections. Then, there is a state w and two sets X, Y such that $X, Y \in N(w)$, but $X \cap Y \notin N(w)$ (recall Lemma 1.2). Since D contains at least two distinct elements, fix $d, c \in D$ such that $d \neq c$. We must construct a model $\mathcal{M} = \langle W, N, D, I \rangle$ based on $\mathcal{F}$ that invalidates the Barcan formula. Let F be a unary predicate symbol that is interpreted as follows:

$$I(F, v) = \begin{cases} \{c\} & \text{iff } v \in X \\ D - \{c\} & \text{iff } v \in Y \\ \emptyset & \text{iff } v \notin X \cup Y \end{cases}$$

Then, we have:

$$[\![F(x)]\!]_{\mathcal{M},\sigma} = \begin{cases} X & \text{iff } \sigma(x) = c \\ Y & \text{iff } \sigma(x) \neq c \end{cases}$$

Since $X, Y \in N(w)$, for any assignment σ, $[\![F(x)]\!]_{\mathcal{M},\sigma} \in N(w)$. Let σ be an assignment. Then, we have $\mathcal{M}, w \models_\sigma \forall x \Box F(x)$. However, $[\![\forall x F(x)]\!]_{\mathcal{M},\sigma} = X \cap Y$; and so, $\mathcal{M}, w \not\models_\sigma \Box \forall x F(x)$.

Suppose, now, that D is infinite and of cardinality κ. Further, suppose that $\mathcal{F}$ is not closed under $\leq \kappa$ intersections. Then, there is a state w and a collection $\{X_i \mid i \in I\}$ where $|I| \leq \kappa$ such that $\bigcap_{i \in I} X_i \notin N(w)$. Since $|I| \leq |D|$, there is a 1-1 function $f : I \to D$. Thus, for each each X_i, there is a unique $c = f(i) \in D$. Denote the element associated with X_i by c^{X_i}. The argument is similar to that in the finite case. Let F be a unary predicate. Define the interpretation of F in such a way that:

$$[\![F(x)]\!]_{\mathcal{M},\sigma} = \begin{cases} X_1 & \text{iff } \sigma(x) \notin \{c^{X_j} \mid j \neq 1, j \in I\} \\ X_i & \text{iff } i \neq 1, i \in I \text{ and } \sigma(x) = c^{X_i} \end{cases}$$

Such an interpretation is possible since each X_i is associated with a unique element c^{X_i} of the domain. Then, for all assignments σ, $[\![F(x)]\!]_{\mathcal{M},\sigma} \in N(w)$. Let σ be an assignment. Then, $\mathcal{M}, w \models_\sigma \forall x \Box F(x)$. However, $[\![\forall x F(x)]\!]_{\mathcal{M},\sigma} = \bigcap_{i \in I} X_i \notin N(w)$; and so, $\mathcal{M}, w \not\models_\sigma \Box \forall x F(x)$. Hence, the Barcan formula is not valid. □

3.2.3 Completeness

The study of first-order modal logics has a long and rich history. It is beyond the scope of this book to survey this extensive literature here. The interested reader is invited to consult Fitting and Mendelsohn (1999), Goldblatt (2011), Hughes and Cresswell (1996), Gabbay et al. (2008), and Garson (2002). There has been much less discussion of *non-normal* first-order modal logics (cf. Arló-Costa 2002; Arló-Costa and Pacuit 2006; Stolpe 2003; Waagbø 1992; Calardo 2013). In this section, I discuss axiomatizations of first-order neighborhood frames with constant domains. Suppose that **L** is a propositional modal logic. Let **FOL** + **L** denote the set of formulas closed under the following rules and axiom schemes (see Hughes and Cresswell 1996, for a discussion):

L   All axiom schemes and rules from **L**.
(All) $\forall x \varphi(x) \to \varphi[y/x]$ is an axiom scheme, where y is free for x in φ.
(Gen) $\frac{\varphi \to \psi}{\varphi \to \forall x \psi}$, where x is not free in φ.

For example, **FOL+E** contains all instances of the axiom schemes (PC), (Dual), and (All) and is closed under the rules (RE), (Gen), and (MP). Given any non-normal or normal propositional modal logic **L**, write $\vdash_{\mathbf{FOL+S}} \varphi$ if $\varphi \in \mathbf{FOL+S}$ (equivalently φ there is a deduction of φ using the above axiom schemas and rules). For simplicity, let **FOL+L+(BF)** denote the set of formulas generated by all axiom schemes and rules of **FOL+S** plus the Barcan formula (BF). The case is similar for **FOL+L+(CBF)**. The first observation is that the converse Barcan formula is derivable in monotonic modal logics.

Exercise 3.7 Prove that $\vdash_{\mathbf{FOL+EM}} \Box \forall x \varphi(x) \to \forall x \Box \varphi(x)$.

The completeness proof for non-normal modal logics with respect to neighborhood semantics combines the canonical model construction from Sect. 2.3.2 with standard techniques for proving completeness for first-order logic (cf. Enderton 2001).

I start by recalling the definitions from Sect. 2.3.2.1. In particular, let $\Gamma \vdash_{\mathbf{FOL+L}} \varphi$ mean that there is a deduction from Γ of φ in the logic **FOL + L**. The definitions of inconsistent, consistent and maximally consistent sets (Definition 2.36) are easily adapted to the first-order modal logic setting. The addition of quantifiers to the modal language does add a complication. The difficulty is that the following set is consistent in any first-order modal logic:

$$\{P(x_1), P(x_2), \ldots, P(x_n), \ldots, \neg \forall x P(x)\}.$$

where the x_i range over all the variables in the language. In order to build a model that satisfies this set, we need to ensure that $\neg \forall x P(x)$ is true. This requires an element of the domain that is not in the interpretation of P. Such an element is called a **witness** for the formula $\neg \forall x P(x)$. This suggest that the maximally consistent sets used to build a canonical model must satisfy an additional property.

Definition 3.12 (∀-*property*) Suppose that Γ is a set of formulas of first-order modal logic. Say that Γ has the ∀-**property** when for each formula $\varphi \in \Gamma$ and each variable x, there is some variable y, called the witness for $\forall x \varphi(x)$, such that $\varphi[y/x] \to \forall x \varphi(x) \in \Gamma$.

As the above example illustrates, we may need to extend the language $\mathcal{L}_1$ in order to find (maximally consistent) sets that have the ∀-property. Suppose that $\mathcal{V}^+$ is a countable set of new variables (i.e., $\mathcal{V}^+ \cap \mathcal{V} = \emptyset$). Suppose that $\mathcal{L}_1^+$ is the first-order modal language generated using the predicate symbols, connectives, quantifiers and modal operators from $\mathcal{L}_1$ and variables from $\mathcal{V} \cup \mathcal{V}^+$. So, $\mathcal{L}_1^+$ extends $\mathcal{L}_1$ with infinitely many new variables. In this extended language, we can adapt Lindenbaum's Lemma (Lemma 2.10) to show that every consistent set of formulas can be extended to be maximally consistent with the ∀-property.

3.2 First-Order Neighborhood Structures

Lemma 3.5 (Lindenbaum's Lemma for First-Order Modal Logic) *Suppose that* **L** *is a first-order modal logic. If Δ is an* **L**-*consistent set of $\mathcal{L}_1$-formulas, then there is a maximally* **L**-*consistent set of $\mathcal{L}_1^+$-formulas Γ with the $\forall$-property such that $\Delta \subseteq \Gamma$, where $\mathcal{L}_1^+$ is the language that extends $\mathcal{L}_1$ with countably many new variables.*

The proof is similar to the proof of the Lindenbaum Lemma for propositional modal logic and is left to the reader.[5] Note that if Γ is an **L**-consistent set of formulas from $\mathcal{L}_1^+$, then Γ is also **L**-consistent with respect to the restricted language $\mathcal{L}_1$. We are now in a position to define a canonical model for first-order modal logic.

Definition 3.13 (*Canonical First-Order Neighborhood Model*) Suppose that **L** is a first-order modal logic. A first-order constant-domain neighborhood model $\mathcal{M}_\mathbf{L} = \langle W_\mathbf{L}, N, D_\mathbf{L}, I_\mathbf{L} \rangle$, where

- $W_\mathbf{L} = \{\Gamma \mid \Gamma$ is a maximally **L**-consistent subset of $\mathcal{L}_1^+$ with the $\forall$-property$\}$;
- $D_\mathbf{L} = \mathcal{V}^+$, where $\mathcal{V}^+$ is the extended set of variables used in the proof of Lemma 3.5; and
- for each $\Gamma \in W_\mathbf{L}$, $\langle x_1, \ldots, x_n \rangle \in I_\mathbf{L}(F, \Gamma)$ iff $F(x_1, \ldots, x_n) \in \Gamma$, where F is an n-ary relation symbol in the language $\mathcal{L}_1^+$.

is **canonical for L** provided that

$$|\varphi|_\mathbf{L} \in N(\Gamma) \text{ iff } \Box\varphi \in \Gamma,$$

where $|\varphi|_\mathbf{L} = \{\Gamma \mid \Gamma$ is a maximally **L**-consistent set with the $\forall$-property, and $\varphi \in \Gamma\}$ is the **proof set** of φ.

The **canonical assignment** is the identity map: $\sigma_\mathbf{L} : \mathcal{V}^+ \to D_\mathbf{L}$, where for all $x \in \mathcal{V}^+$, $\sigma_\mathbf{L}(x) = x$. For example, the **smallest canonical model for L** is $\mathcal{M}_\mathbf{L} = \langle W_\mathbf{L}, N_\mathbf{L}^{min}, D_\mathbf{L}, I_\mathbf{L} \rangle$, where for all $\Gamma \in W_\mathbf{L}$, $N_\mathbf{L}^{min}(\Gamma) = \{|\varphi|_\mathbf{L} \mid \Box\varphi \in \Gamma\}$. Before proving the Truth Lemma, I recall the following standard results about first-order logic (that are easily adapted to the first-order modal setting):

- (**principle of shared variables**) If σ and σ' agree on all free variables in φ, then $\mathcal{M}, w \models_\sigma \varphi$ iff $\mathcal{M}, w \models_{\sigma'} \varphi$.
- (**principle of replacement**) Suppose that σ and σ' are assignments such that for all $u \neq y$, $\sigma'(u) = \sigma(u)$ and $\sigma'(y) = \sigma(x)$. Then, $\mathcal{M}, w \models_\sigma \varphi(x)$ iff $\mathcal{M}, w \models_{\sigma'} \varphi[y/x]$.
- (**principle of alphabetic variants**) If φ and φ' are **alphabetic variants** (the formulas differ only in the names of bound variables), then $\varphi \leftrightarrow \varphi'$ is derivable from the axiom schemes (All), PC and the rules MP and Gen.

---

[5]This is part of the standard toolkit for proving completeness of first-order logic. Consult Enderton (2001) for details. See, also, Hughes and Cresswell (1996), p. 258, for a discussion in the context of first-order modal logic.

Lemma 3.6 (Truth Lemma) *Suppose that* **L** *is a first-order modal logic and* $\mathcal{M} = \langle W_\mathbf{L}, N, D_\mathbf{L}, I_\mathbf{L} \rangle$ *is a canonical model for* **L**. *For each* $\Gamma \in W_\mathbf{L}$ *and formula* $\varphi \in \mathcal{L}_1$,

$$\varphi \in \Gamma \text{ iff } \mathcal{M}_\mathbf{L}, \Gamma \models_{\sigma_\mathbf{L}} \varphi.$$

Proof The proof is by induction on the structure of φ.

The argument for the base case runs as follows:

$$\begin{aligned}
F(x_1, \ldots, x_n) \in \Gamma &\text{ iff } \langle x_1, \ldots, x_n \rangle \in I_\mathbf{L}(F, \Gamma) && \text{(definition of } I_\mathbf{L}\text{)} \\
&\text{ iff } \langle \sigma_\mathbf{L}(x_1), \ldots, \sigma_\mathbf{L}(x_n) \rangle \in I_\mathbf{L}(F, \Gamma) && \text{(definition of } \sigma_\mathbf{L}\text{)} \\
&\text{ iff } \mathcal{M}_\mathbf{L}, \Gamma \models_{\sigma_\mathbf{L}} F(x_1, \ldots, x_n) && \text{(definition of truth)}
\end{aligned}$$

The argument for the Boolean connectives is as usual.

Suppose that $\forall x \varphi(x) \in \Gamma$. Let σ' be any assignment such that $\sigma_\mathbf{L} \sim_x \sigma'$. Then, $\sigma'(x) = y$, where $y \in D_\mathbf{L} = \mathcal{V}^+$. Without loss of generality, we can assume that y is free for x in φ. Otherwise, proceed with the formula $\forall x \varphi'(x)$, where φ' is φ in which all bounded variables y are renamed. Since φ' and φ are alphabetic variants, we have $\mathcal{M}, \Gamma \models_{\sigma_\mathbf{L}} \forall x \varphi(x)$ iff $\mathcal{M}, \Gamma \models_{\sigma_\mathbf{L}} \forall x \varphi'(x)$. Since $\forall x \varphi(x) \in \Gamma$ and y is free for x in φ, using the (All) axiom scheme, we have $\varphi[y/x] \in \Gamma$. By the induction hypothesis, $\mathcal{M}_\mathbf{L}, \Gamma \models_{\sigma_\mathbf{L}} \varphi[y/x]$; and so, by the principle of replacement, $\mathcal{M}_\mathbf{L}, \Gamma \models_{\sigma'} \varphi$. Since σ' is an arbitrary x-variant of $\sigma_\mathbf{L}$, we have $\mathcal{M}_\mathbf{L}, \Gamma \models \forall x \varphi(x)$, as desired.

Suppose that $\forall x \varphi(x) \notin \Gamma$. Then, since Γ is maximal, $\neg \forall x \varphi(x) \in \Gamma$; and so, by the $\forall$-property, there is some variable $y \in \mathcal{V}^+$ such that $\neg \varphi[y/x] \in \Gamma$. That is, $\varphi[y/x] \notin \Gamma$. Thus, by the induction hypothesis, $\mathcal{M}_\mathbf{L}, \Gamma \not\models_{\sigma_\mathbf{L}} \varphi[y/x]$. Thus, if σ' is the x-variant of $\sigma_\mathbf{L}$ with $\sigma'(x) = y$, then, by the principle of replacement, $\mathcal{M}_\mathbf{L}, \Gamma \not\models_{\sigma'} \varphi(x)$. Hence, $\mathcal{M}_\mathbf{L}, \Gamma \not\models_{\sigma_\mathbf{L}} \forall x \varphi(x)$, as desired.

The argument for the modal operator proceeds as in the case for propositional modal logic (Lemma 2.13):

$$\begin{aligned}
\Box \varphi \in \Gamma &\text{ iff } |\varphi|_\mathbf{L} \in N(\Gamma) && \text{(since } \mathcal{M} \text{ is canonical for } \mathbf{L}\text{)} \\
&\text{ iff } [\![\varphi]\!]_{\mathcal{M}, \sigma_\mathbf{L}} \in N(\Gamma) && \text{(induction hypothesis)} \\
&\text{ iff } \mathcal{M}_\mathbf{L}, \Gamma \models_{\sigma_\mathbf{L}} \Box \varphi && \text{(definition of truth)}
\end{aligned}$$

This completes the proof. □

Given the Truth Lemma, completeness for the minimal non-normal first-order modal logic follows from the standard argument (cf. Theorem 2.38).

Theorem 3.14 (Arló-Costa and Pacuit 2006) *The class of all first-order constant-domain neighborhood frames is sound and strongly complete for* **FOL** + **E**.

Completeness for first-order modal logics **FOL** + **L**, where **L** is a non-normal modal logic extending **E** (e.g., **EM**, **EC**, **EMN**, etc.), proceeds as in Sect. 2.3.2. For instance, since we can show that the supplementation of the smallest canonical model for **FOL** + **EM** is a canonical for **FOL** + **EM**, we have:

3.2 First-Order Neighborhood Structures

Theorem 3.15 (Arló-Costa and Pacuit 2006) **FOL + EM** *is sound and complete with respect to the class of supplemented first-order constant-domain neighborhood frames.*

The situation is more complicated when the (converse) Barcan formulas are added to the logic. The first observation is that the axiom scheme M is not derivable in **FOL + E + (CBF)** (the logic that extends **FOL + E** with all instances of (CBF)).

Observation 3.16 $\nvdash_{\mathbf{FOL+E+(CBF)}} \Box(\varphi \wedge \psi) \to (\Box\varphi \wedge \Box\psi)$.

Proof Consider any frame $\mathcal{F} = \langle W, N, D \rangle$ where $|D| = 1$, but N is not closed under supersets. Then, $\Box(\varphi \wedge \psi) \to (\Box\varphi \wedge \Box\psi)$ is not valid on $\mathcal{F}$ (there is a model based on $\mathcal{F}$, a state $w \in W$ and an assignment $\sigma : \mathcal{V} \to D$ such that $\mathcal{M}, w \not\models \Box(\varphi \wedge \psi) \to (\Box\varphi \wedge \Box\psi)$). Note that $\mathcal{F}$ is a frame for **FOL+E+(CBF)** (in particular, all instances of (CBF) are valid on $\mathcal{F}$). Thus, $\nvdash_{\mathbf{FOL+E+(CBF)}} \Box(\varphi \wedge \psi) \to (\Box\varphi \wedge \Box\psi)$. □

This means that **FOL + E + (CBF)** is not strongly complete for the class of first-order constant-domain neighborhood frames that are monotonic. How, then, do we characterize the logic **FOL + E + (CBF)**? The solution is to consider the class of first-order neighborhood frames that are either non-trivial (i.e., $|D| > 1$ and for all $w \in W$, $N(w) \neq \emptyset$) and monotonic or have singleton domains and neighborhood functions that are not monotonic. Then, (CBF) is valid on this class of frames, but the addition of the trivial frames (with $|D| = 1$) with neighborhood functions that are not closed under supersets guarantees that M is not valid. A full discussion of completeness for all non-normal first-order modal logics is beyond the scope of this book. I conclude this presentation of first-order modal logic by discussing the situation with **FOL+K** and **FOL+K+BF**. The first observation is that the Barcan formula is not derivable in **FOL + K**.

Observation 3.17 $\nvdash_{\mathbf{FOL+K}} \forall x \Box \varphi(x) \to \Box \forall x \varphi(x)$.

Proof Suppose that $W = \mathbb{N} \cup \{\infty\}$. For each $n \in \mathbb{N}$, let $[n, \infty] = \{m \mid m \in \mathbb{N}, m \geq n\} \cup \{\infty\}$. For each $i \in W$, let $N(i) = \{[n, \infty] \mid n \in \mathbb{N}\}$. Then, as the reader is invited to check, each $N(i)$ is a filter. Let $D = \{c_0, c_1, \ldots\}$ be a countable set and define an interpretation I such that, for the predicate symbol P, $[\![P(x)]\!]_{\mathcal{M},\sigma} = [n, \infty]$ when $\sigma(x) = c_n$. Since N is a filter, the model $\mathcal{M} = \langle W, N, D, I \rangle$ is a model for **FOL + K**. Suppose that $i \in W$. By construction, we have $[\![P(x)]\!]_{\mathcal{M},\sigma} \in N(i)$, for any assignment σ. Thus, $\mathcal{M}, i \models \forall x \Box P(x)$. However, since $[\![\forall x P(x)]\!]_{\mathcal{M},\sigma} = \{\infty\} \notin N(i)$, $\mathcal{M}, i \not\models \Box \forall x P(x)$. Thus, $\mathcal{M}, i \not\models \forall x \Box P(x) \to \Box \forall x P(x)$. □

Recall from Sect. 2.3.2 that there are two different classes of frames that can be used to characterize the propositional modal logic **K**. The first result that **K** is sound and strongly complete with respect to the class of filters (Theorem 2.42) can be adapted to give a completeness result for **FOL + K**.

Theorem 3.18 **FOL + K** *is sound and strongly complete with respect to the class of constant-domain neighborhood frames that are filters.*

It is well-known that **FOL** + **K** + BF is sound and complete with respect to the class of augmented first-order neighborhood frames. The proof of this result is beyond the scope of this book (see Hughes and Cresswell 1996 and Gabbay et al. 2008 for details).

3.2.3.1 First-Order Topological Models

I conclude my discussion of first-order modal logic by briefly discussing first-order extensions of the topological models from Sect. 1.4.1. A **first-order constant domain topological frame** is a tuple $\langle W, \mathcal{T}, D \rangle$ where $\langle W, \mathcal{T} \rangle$ is a topology (Definition 1.23) and D is a set (to simplify the notation I will call this a **predicate topological frame**). A first-order constant domain topological model (shortened to **first-order predicate model**) is a tuple $\langle W, \mathcal{T}, D, I \rangle$ where $\langle W, \mathcal{T}, D \rangle$ is a predicate topological frame and I is an interpretation of the predicate symbols (for all n-ary predicate symbols F, $I(F, w) \subseteq D^n$).

Recall that the propositional modal logic **S4** extends **K** with the axiom schemes $\Box \varphi \rightarrow \varphi$ and $\Box \varphi \rightarrow \Box \Box \varphi$. It is well-known that **S4** is sound and complete with respect to the class of all topological frames (McKinsey and Tarksi 1944; Rasiowa and Sikorski 1963; Kremer 2013). In the first study of topological models for first-order modal logic, Rasiowa and Sikorski (1963) extended this result to predicate topological frames.

Theorem 3.19 (Rasiowa and Sikorski 1963) *The logic* **FOL** + **S4** *is sound and complete for the class of all predicate topological frames.*

An important line of research is focused on the modal logic of specific topological spaces. Building on the seminal result of McKinsey and Tarski, Rasiowa and Sikorski (1963) proved that **S4** is sound and complete for any topological space that is *dense-in-itself*. Many interesting topological spaces are covered by the Rasiowa and Sikorski result, including the real line (i.e., the usual topology on $\mathbb{R}$), the rational line and the Cantor space. Consult (Kremer 2013; Bezhanishvili and Gehrke 2005; Lando 2012; Mints and Zhang 2005) for further refinements and alternative proofs of these fundamental theorems. This raises an interesting question about whether **FOL** + **S4** is complete for the corresponding predicate topological spaces. I start with a negative result from Kremer (2014). Say that a first-order modal logic **FOL** + **L** is complete for a topology $\langle W, \mathcal{T} \rangle$ provided that **FOL** + **L** is complete for the class of predicate topological frames $\langle W, \mathcal{T}, D \rangle$, where D is any domain.

Theorem 3.20 (Kremer 2014) *The logic* **FOL** + **S4** *is not complete for the real line (i.e., $\langle \mathbb{R}, \mathcal{T}_\mathbb{R} \rangle$ where $\mathcal{T}_\mathbb{R}$ is the usual topology).*

Proof Let φ be the formula:

$$\forall x \Box F(x) \wedge \Box \forall x (\Box F(x) \vee \Box \neg F(x)) \rightarrow \Box \forall x F(x),$$

where F is a unary predicate. We will show that

3.2 First-Order Neighborhood Structures 115

1. φ is valid on any predicate topological frame $\langle \mathbb{R}, \mathcal{T}_\mathbb{R}, D\rangle$, where $\langle \mathbb{R}, \mathcal{T}_\mathbb{R}\rangle$ is the usual topology on $\mathbb{R}$; and
2. $\nvdash_{\mathbf{FOL+S4}} \varphi$.

To prove item 1, let $\mathcal{M} = \langle \mathbb{R}, \mathcal{T}_\mathbb{R}, D, V\rangle$ be a first-order topological model based on $\langle \mathbb{R}, \mathcal{T}_\mathbb{R}, D\rangle$. I start with an observation about the topology $\langle \mathbb{R}, \mathcal{T}_\mathbb{R}\rangle$. Suppose that $I \subseteq \mathbb{R}$ is an interval. That is, $I = (r_1, r_2) = \{r \in \mathbb{R} \mid r_1 < r < r_2\}$. Then, there are no open sets $O_1, O_2 \in \mathcal{T}_\mathbb{R}$ such that $O_1 \cap O_2 = \emptyset$ and $I = O_1 \cup O_2$. That is, no interval is the disjoint union of open sets. Topological spaces with this property are said to be **locally connected**.

Let $r \in \mathbb{R}$ and $\sigma : \mathcal{V} \to D$ be an assignment. Suppose that $\mathcal{M}, r \nvDash_\sigma \varphi$. Then,

(1) $\mathcal{M}, r \vDash_\sigma \forall x \Box F(x)$;
(2) $\mathcal{M}, r \vDash_\sigma \Box \forall x(\Box F(x) \vee \Box \neg F(x))$; and
(3) $\mathcal{M}, r \nvDash_\sigma \Box \forall x F(x)$.

By (2), there is an $O \in \mathcal{T}_\mathbb{R}$ such that $r \in O$ and for all σ', if $\sigma \sim_x \sigma'$, then

(4) $\quad$ for all σ' if $\sigma \sim_x \sigma'$, then $O \subseteq [\![\Box F(x) \vee \Box \neg F(x)]\!]_{\mathcal{M},\sigma'}$.

By (3), there is an assignment σ'' such that $\sigma \sim_x \sigma''$ and

(5) $\quad O \cap [\![\neg F(x)]\!]_{\mathcal{M},\sigma''} \neq \emptyset$

Since $\sigma \sim_x \sigma''$, by (4), we have that

(6) $\quad O \subseteq [\![\Box F(x) \vee \Box \neg F(x)]\!]_{\mathcal{M},\sigma''} = [\![\Box F(x)]\!]_{\mathcal{M},\sigma''} \cup [\![\Box \neg F(x)]\!]_{\mathcal{M},\sigma''}$

Since $r \in O \in \mathcal{T}_\mathbb{R}$, there is an interval I such that $r \in I \subseteq O$. Define the sets:

$$O_1 = I \cap [\![\Box F(x)]\!]_{\mathcal{M},\sigma''} \qquad O_2 = I \cap [\![\Box \neg F(x)]\!]_{\mathcal{M},\sigma''}$$

Then, O_1 and O_2 are both open sets (they both are finite intersections of open sets). Furthermore, by (6) and that $I \subseteq O$, we have that $I = O_1 \cup O_2$. By (1), $r \in [\![\Box F(x)]\!]_{\mathcal{M},\sigma''}$. Hence, $r_1 \in I \cap [\![\Box F(x)]\!]_{\mathcal{M},\sigma''}$; and so, $O_1 \neq \emptyset$. By (5) and that $I \subseteq O$, there is a $t \in \mathbb{R}$ such that $t \in I \cap [\![\neg F(x)]\!]_{\mathcal{M},\sigma''}$. Since $[\![\Box F(x)]\!]_{\mathcal{M},\sigma''} \subseteq [\![F(x)]\!]_{\mathcal{M},\sigma''}$, we have $t \notin [\![\Box F(x)]\!]_{\mathcal{M},\sigma''}$. By (6), $t \in [\![\Box \neg F(x)]\!]_{\mathcal{M},\sigma''}$. Thus, $t \in I \cap [\![\Box \neg F(x)]\!]_{\mathcal{M},\sigma''}$; and so $O_2 \neq \emptyset$. Hence, the interval I is the disjoint union of open sets. This is contradicts the above observation that $\langle \mathbb{R}, \mathcal{T}_\mathbb{R}\rangle$ is locally connected. Thus, φ is valid on any predicate topological frame $\langle \mathbb{R}, \mathcal{T}_\mathbb{R}, D\rangle$.

To prove item 2, we show that φ is not valid on a class of frames for which **FOL + S4** is known to be complete. A **variable domain first-order relational frame** is a tuple $\langle W, R, \mathcal{D}, I\rangle$, where $W \neq \emptyset$; $R \subseteq W \times W$; $\mathcal{D}$ is a family of sets indexed by states W (write $\mathcal{D}_w$ for the domain associated with w); and I is a variable domain interpretation function: for each state w and n-ary predicate F, $I(w, F) \subseteq \mathcal{D}_w^n$. Formulas from the first-order modal language $\mathcal{L}_1$ are interpreted at states w and assignments $\sigma : \mathcal{V} \to \bigcup_{w \in W} \mathcal{D}_w$. The definition of truth for atomic formulas and the Boolean connectives is similar to Definition 3.7. Truth for the universal quantifier and the modal operator is defined as follows:

- $\mathcal{M}, w \models_\sigma \forall x \varphi(x)$ for all σ', if $\sigma \sim_x \sigma'$ and $\sigma'(x) \in \mathcal{D}_w$, then $\mathcal{M}, w \models_{\sigma'} \varphi(x)$
- $\mathcal{M}, w \models_\sigma \Box \varphi$ for all v, if $w \, R \, v$, then $\mathcal{M}, v \models_\sigma \varphi$

It is known that **FOL + S4** is sound and complete with respect to variable domain first-order relational models with a reflexive and transitive relation (Hughes and Cresswell 1996). It is not hard to find a model based on such a frame that makes φ false. Let $W = \{w, v\}$, $\mathcal{D}_w = \mathcal{D}_v = \{a, b\}$, $R = \{(w, w), (w, v), (v, v)\}$ and $I(w, F) = I(v, F) = \{a\}$. Let $\sigma_1(x) = a$ and $\sigma_2(x) = b$. Then, $\mathcal{M}, w \models_\sigma \forall x \Box F(x)$ since the only element in the domain $\mathcal{D}_w$ is a and $a \in I(v, F) \cap I(w, F)$. To see that $\mathcal{M}, w \models_\sigma \Box \forall x (\Box F(x) \vee \Box \neg F(x))$, note that a is the only element of $\mathcal{D}_w$, so clearly $\mathcal{M}, w \models_\sigma \forall x (\Box F(x) \vee \Box \neg F(x))$. Furthermore, since v is the only state accessible from v, $\mathcal{M}, v \models_{\sigma'} \Box F(x) \vee \Box \neg F(x)$ for any assignment σ'. Finally, $\mathcal{M}, w \not\models_\sigma \Box \forall x F(x)$ since $w \, R \, v$ and $\mathcal{M}, v \not\models_{\sigma'} F(x)$ when $\sigma'(x) = b$. Thus, $\forall x \Box F(x) \wedge \Box \forall x (\Box F(x) \vee \Box \neg F(x)) \to \Box \forall x F(x)$ is not valid on these variable domain relational frames, which implies that $\not\vdash_{\mathbf{FOL+S4}} \varphi$. $\square$

Interestingly, (Kremer, 2014) showed that **FOL + S4** is sound and complete with respect to the *rational line*. The rational line is the *subspace topology induced* by $\mathcal{T}_\mathbb{R}$. That is, it is the topology $\langle \mathbb{Q}, \mathcal{T}_\mathbb{Q} \rangle$, where $\mathbb{Q} \subseteq \mathbb{R}$ is the set of rational numbers and $\mathcal{T}_\mathbb{Q} = \{U \cap \mathbb{Q} \mid U \in \mathcal{T}_\mathbb{R}\}$. I report the theorem here, and refer the reader to (Kremer 2014, Theorem 6.1) for the intricate proof.

Theorem 3.21 (Kremer 2014) **FOL + S4** *is sound and complete for* $\langle \mathbb{Q}, \mathcal{T}_\mathbb{Q}, D \rangle$, *where D is countably infinite.*

My discussion of first-order modal logic on neighborhood structures was simplified in two ways. First, the first-order modal language $\mathcal{L}_1$ does not include constants, function symbols or equality. Second, I restricted attention to constant domain models. It is also common to consider **variable domain models** in which different domains are assigned to different states (cf. the proof of item 2 in Theorem 3.20). Consult Gabbay et al. (2008), Awodey and Kishida (2008), Kishida (2011); Calardo (2013), and Kremer (2014) for further results about first-order modal logic on neighborhood structures.

3.3 Common Belief on Neighborhood Structures

The game theory and epistemic logic literature contains many notions of *group* knowledge and belief. These notions have played a fundamental role in the analysis of distributed algorithms (Halpern and Moses 1990), social interactions (Chwe 2001)

3.3 Common Belief on Neighborhood Structures

and political institutions (List 2014). It is beyond the scope of this section to discuss all of these concepts (see Vanderschraaf and Sillari 2014, for an in-depth discussion of this literature).[6] In this section, I introduce multi-agent neighborhood models and show how to define various notions of group beliefs in this setting.

Suppose that $\mathcal{A} = \{1, \ldots, n\}$ is a finite set of agents. A **multi-agent modal language** is defined in the obvious way. Suppose that At is a set of atomic propositions, and let $\mathcal{L}^n$ contains all formulas generated from the following grammar:

$$p \mid \neg \varphi \mid (\varphi \wedge \psi) \mid \Box_i \varphi$$

where $p \in \mathsf{At}$ and $i \in \mathcal{A}$. Additional Boolean connectives ($\vee, \rightarrow, \leftrightarrow$) are defined as usual, and for each $i \in \mathcal{A}$, let $\Diamond_i \varphi$ be $\neg \Box_i \neg \varphi$. In this section, the intended interpretation of $\Box_i \varphi$ is that "agent i *believes* that φ".

A **multi-agent neighborhood frame** is a tuple $\langle W, \{N_i\}_{i \in \mathcal{A}} \rangle$, where for each $i \in \mathcal{A}$, $N_i : W \rightarrow \wp(\wp(W))$ is a neighborhood function. A **multi-agent neighborhood model** is a tuple $\mathcal{M} = \langle W, \{N_i\}_{i \in \mathcal{A}}, V \rangle$, where $\langle W, \{N_i\}_{i \in \mathcal{A}} \rangle$ is a multi-agent neighborhood frame and $V : \mathsf{At} \rightarrow \wp(W)$ is a valuation function. Truth of formulas $\varphi \in \mathcal{L}^n$ at states in a multi-agent neighborhood model $\mathcal{M} = \langle W, \{N_i\}_{i \in \mathcal{A}}, V \rangle$ is defined as in Definition 1.12. Truth of the indexed modal formulas runs as follows:

$$\mathcal{M}, w \models \Box_i \varphi \text{ iff } [\![\varphi]\!]_{\mathcal{M}} \in N_i(w).$$

In the remainder of this section, I discuss various notions of group beliefs in multi-agent neighborhood models.

Everyone Believes

Suppose that $G \subseteq \mathcal{A}$ is a non-empty set of agents, and $\mathcal{M} = \langle W, \{N_i\}_{i \in \mathcal{A}}, V \rangle$ is a multi-agent neighborhood model. Let $N_G : W \rightarrow \wp(\wp(W))$ be a neighborhood function where, for all $w \in W$, $N_G(w) = \bigcap_{i \in G} N_i(w)$. Thus, $N_G(w)$ contains all sets that are neighborhoods for each agent $i \in G$. Extend the language $\mathcal{L}^n$ with operators $\Box_G$ for each $G \subseteq \mathcal{A}$. Truth of $\Box_G \varphi$ is defined as follows:

$$\mathcal{M}, w \models \Box_G \varphi \text{ iff } [\![\varphi]\!]_{\mathcal{M}} \in N_G(w).$$

Thus, $\Box_G \varphi$ is true if every agent in G believes that φ. Since there are only finitely many agents, we can define $\Box_G \varphi$ as $\bigwedge_{i \in G} \Box_i \varphi$. Many of the properties of neighborhood functions discussed in Sect. 1.1 "lift" to the everyone believes neighborhood function.

Exercise 3.8 1. If, for all $i \in G$, N_i is monotonic, then N_G is monotonic.
2. If, for all $i \in G$, N_i is augmented, then N_G is augmented.
3. Is there a property P of neighborhood functions such that for each $i \in G$, N_i has property P, but N_G does not have property P?

---

[6]The textbooks Fagin et al. (1995) and van Benthem (2011) also provide illuminating discussions of key logical issues.

Exercise 3.9 Prove that (RE) plus the axiom

$$\Box_G \varphi \leftrightarrow \bigwedge_{i \in G} \Box_i \varphi$$

is sound and complete for all multi-agent neighborhood frames in which N_G is defined as above.

Distributed Belief

The neighborhood function N_G for a set of agents G assigns to each state w the sets that all agents in G at w believe. A more interesting operation on neighborhood functions is to *aggregate* the agents' beliefs. Suppose that $\mathcal{X}$ and $\mathcal{Y}$ are two collections of subsets of W. We can form a new collection of sets by intersecting each of the elements of $\mathcal{X}$ and $\mathcal{Y}$:

$$\mathcal{X} \sqcap \mathcal{Y} = \{Z \mid Z = X \cap Y, \text{ for some } X \in \mathcal{X} \text{ and } Y \in \mathcal{Y}\}.$$

Using the above operation, we can define a neighborhood function that represents an *aggregation* of a group's beliefs. Suppose that $\mathcal{M} = \langle W, \{N_i\}_{i \in \mathcal{A}}, V \rangle$ is a multi-agent neighborhood model. For each $\emptyset \neq G \subseteq \mathcal{A}$, define the aggregate neighborhood function $N_G^\sqcap : W \to \wp(\wp(W))$ as follows. For each $w \in W$,

$$N_G^\sqcap(w) = \sqcap_{i \in G} N_i(w).$$

To reason about this operation, add a modality $[\sqcap]_G$ to the language $\mathcal{L}^n$. The interpretation of $[\sqcap]_G \varphi$ is as follows:

$$\mathcal{M}, w \models [\sqcap]_G \varphi \text{ iff } [\![\varphi]\!]_\mathcal{M} \in N_G^\sqcap(w).$$

The formula $[\sqcap]_G \varphi$ corresponds to what is called **distributed belief** in the epistemic logic literature (cf. Halpern and Moses 1990; Roelofsen 2007; van der Hoek et al. 1999). The distributed beliefs of a group represent what the agents would believe if they shared everything that they believe. The following example illustrates this. Suppose that $\mathcal{A} = \{1, 2\}$, and $\mathcal{M} = \langle W, \{N_1, N_2\}, V \rangle$ is a multi-agent neighborhood model with $W = \{w_1, w_2, w_3, w_4\}$, $V(p) = \{w_1, w_3\}$ and $V(q) = \{w_1, w_2\}$. The agents' neighborhoods at state w_1 are (the neighborhoods at other states do not matter for this example):

- $N_1(w_1) = \{\{w_1, w_3\}, \{w_1, w_2, w_3\}, \{w_1, w_3, w_4\}, W\}$, and
- $N_2(w_1) = \{\{w_1, w_4\}, \{w_1, w_2, w_4\}, \{w_1, w_3, w_4\}, W\}$.

Since $[\![p]\!]_\mathcal{M} = \{w_1, w_3\}$ and $[\![p \to q]\!]_\mathcal{M} = \{w_1, w_2, w_4\}$, we have that $\mathcal{M}, w_1 \models \Box_1 p$ and $\mathcal{M}, w_1 \models \Box_2(p \to q)$. However, neither agent believes that q: $\mathcal{M}, w_1 \not\models \Box_1 q$ and $\mathcal{M}, w_1 \not\models \Box_2 q$. There is distributed belief of q ($\mathcal{M}, w_1 \models [\sqcap]_{\{1,2\}} q$). To see this, note that $N_1(w_1) \sqcap N_2(w_1)$ is the set

$$\{\{w_1\}, \{w_1, w_2\}, \{w_1, w_3\}, \{w_1, w_4\}, \{w_1, w_2, w_3\}, \{w_1, w_3, w_4\}, \{w_1, w_2, w_4\}, W\}.$$

3.3 Common Belief on Neighborhood Structures

Note that $N_G^\sqcap(w)$ may contain $\emptyset$ even when $\emptyset \notin N_i(w)$ for each $i \in G$. This would happen if the agents' beliefs were inconsistent (the same is true for the distributed belief operator discussed in Halpern and Moses (1990), Roelofsen (2007). The above example shows that there may be formulas that are distributed beliefs but are not believed by any of the agents in the group. That is, for a group of agents G with $i \in G$, the formula $[\sqcap]_G \varphi \to \square_i \varphi$ (for $i \in G$) is not necessarily valid. The converse is valid given a natural assumption about the neighborhoods.

Exercise 3.10 1. Prove that in multi-agent neighborhood models in which each neighborhood contains the unit (for all $i \in \mathcal{A}$, for all $w \in W$, $W \in N_i(w)$), the formula $\bigwedge_{i \in G}(\square_i \varphi \to [\sqcap]_G \varphi)$ is valid.
2. Show that the formula from part 1. is not necessarily valid if the neighborhoods do not contain the unit.

I conclude this short subsection with an exercise that demonstrates that the above operator is, in fact, the appropriate generalization of distributed belief.

Exercise 3.11 Suppose that for all $i \in G$, N_i is augmented. Prove that $N_G^\sqcap$ is augmented, and that $R_{N_G^\sqcap} = \bigcap_{i \in G} R_{N_i}$ (cf. Lemma 2.3).

Common Belief

Distributed belief describes what the group believes after everyone in the group shares everything that they believe. At the opposite end of the spectrum is *common belief*. This is a group belief that is completely transparent to everyone in the group (without any communication). That is, φ is commonly believed if everyone believes that φ, everyone believes that everyone believes that φ, everyone believes that everyone believes that everyone believes that φ, and so on *ad infinitum*.[7] Instead, I focus on the definition of common belief in weak systems of modal logic (Lismont and Mongin 1994a, b, 2003; Heifetz 1996).

In the remainder of this section, I restrict attention to multi-agent neighborhood frames that are monotonic. That is, the neighborhood function $N_i : W \to \wp(\wp(W))$ for each agent $i \in \mathcal{A}$ satisfies the monotonicity property: For all $w \in W$ and sets $X, Y \subseteq W$, if $X \subseteq Y$ and $X \in N_i(w)$, then $Y \in N_i(w)$. Crucially, the neighborhood functions may not be closed under intersection. It turns out that there are two definitions of common belief in this setting. Both definitions use the following notion.

Definition 3.22 (*Evident Beliefs*) Suppose that $\mathcal{M} = \langle W, \{N_i\}_{i \in \mathcal{A}}, V \rangle$ is a multi-agent neighborhood model. A set $X \subseteq W$ is i-**evident** provided that $X \subseteq \{w \mid X \in N_i(w)\}$. A set X is G-**evident** if it is i-evident for all $i \in G$. To simplify the notation, we say that X is **evident** if it is $\mathcal{A}$-evident.

---

[7]There are rich literatures in logic (Halpern and Moses 1990; Heifetz 1999; van Benthem and Sarenac 2004; Barwise 1987), philosophy (Cubitt and Sugden 2003; Lewis 1969), and game theory (Geanakoplos 1995; Monderer and Samet 1989; Pacuit and Roy 2015) focused on common belief. It is beyond the scope of this book to discuss this literature.

Recall from Sect. 1.2.1 that if $N : W \to \wp(\wp(W))$ is a neighborhood function, then $m_N : \wp(W) \to \wp(W)$ is the function, where for all $X \subseteq W$, $m_N(X) = \{w \mid X \in N(w)\}$. To simplify notation, for $G \subseteq \mathcal{A}$, I write m_G instead of m_{N_G}. Thus, $m_G(X)$ is the set of all states in which every agent in G believes X; and X is G-evident if $X \subseteq m_G(X)$.

The first definition of common belief is a non-probabilistic version of the common belief operator from Monderer and Samet (1989).

Definition 3.23 (*Common Belief, version 1*) Let $\mathcal{M} = \langle W, \{N_i\}_{i \in \mathcal{A}}, V \rangle$ be a monotonic multi-agent neighborhood model. For each $G \subseteq \mathcal{A}$, define a function $C_G^1 : \wp(W) \to \wp(W)$ as follows:

$$C_G^1(Y) = \{w \mid \text{there is a } G\text{-evident set } X \text{ such that } w \in X \text{ and } X \subseteq m_G(Y)\}.$$

Then, φ is commonly believed at a state w if there is some evident set containing w that implies that everyone believes that φ. It will be useful to work with the following characterization of the common belief operator:

$$(*) \quad C_G^1(Y) = \bigcup \{X \mid X \subseteq m_G(X) \cap m_G(Y)\}.$$

To reason about this operator, I extend the language $\mathcal{L}^n$ with a modal operator $\square_G^*$ with the following truth clause:

$$\mathcal{M}, w \models \square_G^* \varphi \text{ iff } w \in C_G^1(\llbracket \varphi \rrbracket_{\mathcal{M}}).$$

Let $\mathbf{EM}_n^C$ be the smallest set of formulas that contains all tautologies in the above language; contains all instances of the following axiom schemes; and is closed under the following rules:

(RM$_n$) $\quad \dfrac{\varphi \to \psi}{\square_i \varphi \to \square_i \psi}$

(CB) $\quad \square_G^* \varphi \to \square_G \varphi$

(FP1) $\quad \square_G^* \varphi \to \square_G \square_G^* \varphi$

(FP2) $\quad \dfrac{(\varphi \to \square_G \varphi) \wedge (\varphi \to \square_G \psi)}{\varphi \to \square_G^* \psi}$

It a simple (and illuminating!) exercise to show that $\mathbf{EM}_n^C$ is sound for the class of monotonic multi-agent neighborhood models.

Proposition 3.4 *The logic* $\mathbf{EM}_n^C$ *is sound for the class of monotonic multi-agent neighborhood models.*

Proof The rule RM_n is the generalization of the monotonicity rules that we already know is valid on the class of monotonic neighborhood frames (cf. Theorem 2.14). The proof that the rule (FP2) is valid is a direct consequence of the characterization of the

3.3 Common Belief on Neighborhood Structures

common belief operator given in (∗). We conclude by proving that (CB) and (FP1) are valid. Suppose that $\mathcal{M} = \langle W, \{N_i\}_{i \in \mathcal{A}}, V\rangle$ is a monotonic multi-agent neighborhood model and $w \in W$. Suppose that $\mathcal{M}, w \models \Box_G^* \varphi$. Then, there is a G-evident set $X \subseteq W$ such that $w \in X$ and $X \subseteq m_G(\llbracket \varphi \rrbracket_\mathcal{M})$. Thus, $X \subseteq m_G(X) \cap m_G(\llbracket \varphi \rrbracket_\mathcal{M})$. Since $m_G(\llbracket \varphi \rrbracket_\mathcal{M}) = \llbracket \Box_G \varphi \rrbracket_\mathcal{M}$ and $w \in X$, we have that $\mathcal{M}, w \models \Box_G \varphi$. Thus, (CB) is valid. Furthermore, by (∗), we have $X \subseteq C_G^1(\llbracket \varphi \rrbracket_\mathcal{M}) = \llbracket \Box_G^* \varphi \rrbracket_\mathcal{M}$. Now, X is G-evident and, by monotonicity, we have that $w \in X \subseteq m_G(X) \subseteq m_G(\llbracket \Box_G^* \varphi \rrbracket_\mathcal{M}) = \llbracket \Box_G \Box_G^* \varphi \rrbracket_\mathcal{M}$, which implies that $\mathcal{M}, w \models \Box_G \Box_G^* \varphi$. Thus, (FP1) is valid. □

Exercise 3.12 Prove that if $\vdash_{\mathbf{EM}_n^c} \varphi \to \psi$, then $\vdash_{\mathbf{EM}_n^c} \Box_G^* \varphi \to \Box_G^* \psi$.

The discussion so far has focused on the use of evident sets to define common belief. This is the so-called *fixed-point* definition of common belief (cf. Heifetz 1999; Barwise 1987). How is this fixed-point definition related to the iterative characterization of common belief that was referenced in the first paragraph of this subsection? For each $k \in \mathbb{N}$, let $\Box_G^k \varphi$ be the formula that consists of a sequence of $k + 1$ everyone believes modalities. Formally, let $\Box_G^0 \varphi = \Box_G \varphi$ and for $k > 0$, $\Box_G^k \varphi = \Box_G(\Box_G^{k-1} \varphi)$. The first observation is that for each $k \in \mathbb{N}$, $\Box_G^* \varphi \to \Box_G^k$ is valid. When $k = 0$, this is just the validity of axiom (CB). The proof that for each $k > 0$, $\Box_G^* \varphi \to \Box_G^k$ is valid proceeds by induction on k. The key observation is that if $w \in C_G^1(\llbracket \varphi \rrbracket_\mathcal{M})$, then there is a G-evident set X such that $w \in X \subseteq m_G(X) \cap m_G(\llbracket \varphi \rrbracket_\mathcal{M})$. By monotonicity, we have that $m_G(X) \subseteq m_G(m_G(\llbracket \varphi \rrbracket_\mathcal{M}))$. Since X is G-evident, we have that $w \in X \subseteq m_G(X) \subseteq m_G(m_G(\llbracket \varphi \rrbracket_\mathcal{M}))$. We can continue in this manner to show that $\mathcal{M}, w \models \Box_G^k \varphi$ for any $k > 0$. I leave it to the reader to show that the above argument can be reproduced inside the logic.

Exercise 3.13 Prove that $\vdash_{\mathbf{EM}_n^c} \Box_G^* \varphi \to \Box_G^k \varphi$ for all $k \in \mathbb{N}$.

Thus, the truth of $\Box_G^* \varphi$ implies that each of the formulas in the set $\{\Box_G^k \varphi \mid k \in \mathbb{N}\}$ are true. That is, common belief in φ implies that everyone believes that φ and everyone believes that everyone believes that φ, and so on. Of course, this cannot be expressed in our language since the language contains only finite conjunctions. Setting aside questions about expressing infinite conjunctions in the language, there is still a question about whether we can conclude that $\Box_G^* \varphi$ is true if we know that conjunctions of formulas from $\{\Box_G^k \varphi \mid k \in \mathbb{N}\}$ are true. There are two key issues that arise when the neighborhoods are not closed under intersections. The first issue is that there is more than one way to form intersections from the set $\{\llbracket \Box_G^k \varphi \rrbracket_\mathcal{M} \mid k \in \mathbb{N}\}$. Since the neighborhoods are not closed under intersections, it is not hard to find a set X such that:

$$m_G(X \cap m_G(X)) \neq m_G(X) \cap m_G(m_G(X)).$$

Each side of this inequality can be used to form infinite conjunctions of everyone believes statements. It turns out that the right-hand side corresponds to the first version of common belief from Definition 3.23. For a set $X \subseteq W$, define X_α, where α is any infinite cardinal, by transfinite induction:

$$X_0 = m_G(X)$$
$$X_\alpha = \bigcap_{\beta<\alpha} X_\beta \cap m_G(\bigcap_{\beta<\alpha} X_\beta)$$

Note that the above sequence of sets is decreasing. Thus, for any set W, there must be some μ such that $X_\mu = X_{\mu+1}$. This brings us to the second issue. In general, μ may not be $\aleph_0$ (the first countable cardinal). That is, $\bigcap_{n\in\mathbb{N}} X_n$ may contain states that are not in $m_G(\bigcap_{n\in\mathbb{N}} X_n)$. This is very different from the situation with common belief (knowledge) defined on relational structures. It is not hard to see that on relational structures, for any set X, X_α must stabilize at $\aleph_0$.

Exercise 3.14 Suppose that $\mathcal{M} = \langle W, \{N_i\}_{i\in\mathcal{A}}, V\rangle$ is a multi-agent neighborhood model in which each N_i is augmented. For any $X \subseteq W$, prove that X_α stabilizes when $\alpha = \aleph_0$.

The main result of this short section is that the above definition of (possibly transfinite) iterations of the everyone believes operator corresponds to the definition of common belief from Definition 3.23.

Proposition 3.5 *Suppose that $\mathcal{M} = \langle W, \{N_i\}_{i\in\mathcal{A}}, V\rangle$ is a monotonic multi-agent neighborhood model. For any set $Y \subseteq W$ and set $G \subseteq \mathcal{A}$, $C_G^1(Y) = Y_\mu$, where μ is the least cardinal such that $Y_\mu = Y_{\mu+1}$.*

Proof Suppose that μ is the smallest cardinal such that $Y_\mu = Y_{\mu+1}$. We must show that $C_G^1(Y) = Y_\mu$. We proceed in two steps.

Claim 1: $Y_\mu \subseteq C_G^1(Y)$. Suppose that $w \in Y_\mu$. Since $Y_0, Y_1, \ldots, Y_\mu$ is a decreasing sequence, we have $\bigcap_{\beta<\mu+1} Y_\beta = Y_\mu$. Thus, $Y_{\mu+1} = \bigcap_{\beta<\mu+1} Y_\beta \cap m_G(\bigcap_{\beta<\mu+1} Y_\beta) = Y_\mu \cap m_G(Y_\mu)$. Since $Y_\mu = Y_{\mu+1} = Y_\mu \cap m_G(Y_\mu)$, we have $Y_\mu \subseteq m_G(Y_\mu)$. Thus, Y_μ is evident. Furthermore, $Y_\mu \subseteq Y_0 = m_G(Y)$. Thus, there is an evident set containing w that implies that everyone believes Y. Hence, $w \in C_G^1(Y)$.

Claim 2: $C_G^1(Y) \subseteq Y_\mu$. Suppose that $w \in C_G^1(Y)$. Then, there is a $X \subseteq W$ such that $X \subseteq m_G(X)$, $w \in X$ and $X \subseteq m_G(Y)$. We prove by transfinite induction on α that $X \subseteq Y_\alpha$. The base case is $X \subseteq Y_0 = m_G(Y)$. This follows by assumption. Suppose that for all $\beta < \alpha$, $X \subseteq Y_\beta$. We must show that $X \subseteq Y_\alpha$. By the induction hypothesis, we have $X \subseteq \bigcap_{\beta<\alpha} Y_\beta$. By monotonicity and the fact that X is evident, we have,

$$X \subseteq m_G(X) \subseteq m_G(\bigcap_{\beta<\alpha} Y_\beta).$$

Hence, $X \subseteq \bigcap_{\beta<\alpha} Y_\beta \cap m_G(\bigcap_{\beta<\alpha} Y_\beta) = Y_\alpha$. □

The second version of the definition of common belief is from Lismont and Mongin (1994a, b).

Definition 3.24 (*Common Belief, version 2*) Let $\mathcal{M} = \langle W, \{N_i\}_{i\in\mathcal{A}}, V\rangle$ be a monotonic multi-agent neighborhood model. For each $G \subseteq \mathcal{A}$, define a function $C_G^2 : \wp(W) \to \wp(W)$ as follows:

$$C_G^2(Y) = \{w \mid \text{there is a } G\text{-evident set } X \text{ such that } w \in m_G(X) \text{ and } X \subseteq Y\}.$$

3.3 Common Belief on Neighborhood Structures

This definition of common belief corresponds to the second way to form conjunctions of everyone believes statements. For a set $X \subseteq W$, define $\hat{X}_\alpha$, where α is any infinite cardinal, by transfinite induction:

$$\hat{X}_0 = m_G(X)$$
$$\hat{X}_\alpha = m_G(X \cap \bigcap_{\beta < \alpha} \hat{X}_\beta)$$

This version of common beliefs satisfies many of the same properties as the first version.

Exercise 3.15 Suppose that $\mathcal{M} = \langle W, \{N_i\}_{i \in \mathcal{A}}, V \rangle$ is a monotonic multi-agent neighborhood model. Let $\square_G^\star$ be a modal operator with the truth clause $\mathcal{M}, w \models \square_G^\star \varphi$ iff $w \in C_G^2(\llbracket \varphi \rrbracket_{\mathcal{M}})$.

1. Prove that for any set $Y \subseteq W$ and set $G \subseteq \mathcal{A}$, $C_G^2(Y) = \hat{Y}_\mu$, where μ is the least cardinal such that $\hat{Y}_\mu = \hat{Y}_{\mu+1}$.
2. Prove that $\square_G^\star \varphi \to \square_G \varphi$ and $\square_G^\star \varphi \to \square_G \square_G^\star \varphi$ are valid.
3. Find a variant of (FP2) that is valid for the $\square_G^\star$ modality (cf. Lismont and Mongin 1994a).

I conclude this subsection by clarifying the relationship between the two versions of common belief.

Proposition 3.6 Suppose that $\mathcal{M} = \langle W, \{N_i\}_{i \in \mathcal{A}}, V \rangle$ is a monotonic multi-agent neighborhood model. For any set $Y \subseteq W$, $C_G^2(Y) \subseteq C_G^1(Y)$.

Proof Suppose that $w \in C^2(Y)$. Then, there exists an evident set $X \subseteq W$ such that $w \in m_G(X)$ and $X \subseteq Y$. Thus, by monotonicity, we have $m_G(X) \subseteq m_G(m_G(X))$ and $m_G(X) \subseteq m_G(Y)$. Hence, $m_G(X)$ is an evident set containing w, which implies that everyone believes that Y. Thus, $w \in C^1(Y)$. □

The converse of Proposition 3.6 is not true. That is, there is a model with a set Y such that $C_G^1(Y) \not\subseteq C_G^2(Y)$. Suppose that $W = \{w, x, y\}$ and all agents have the same neighborhood function $N : W \to \wp(\wp(W))$:

- $N(w) = \{\{w\}, \{v\}, \{x\}, \{w, v\}, \{w, x\}, \{v, x\}, \{w, v, x\}\}$
- $N(v) = \{\{w, v\}, \{w, x\}, \{v, x\}, \{w, v, x\}\}$
- $N(x) = \{\{w, v, x\}\}$

Then, $X = \{w\}$ is evident: $m_\mathcal{A}(\{w\}) = \{w\}$. This implies that $w \in C_\mathcal{A}^1(\{v, x\})$ since $w \in X$, X is evident and $X \subseteq m_\mathcal{A}(\{v, x\}) = \{w, v\}$. However, $w \notin C_\mathcal{A}^2(\{v, x\})$. To see this, note that there are no subsets of $\{v, x\}$ that are evident ($m_\mathcal{A}(\{v\}) = m_\mathcal{A}(\{x\}) = \{w\}$, $m_\mathcal{A}(\{v, x\}) = \{w, v\}$).[8]

There is much more to say about common belief on neighborhood structures. For instance, there is the question of completeness for languages with each of the above

---

[8] Technically, the emptyset $\emptyset$ is evident since $\emptyset = m_G(\emptyset)$; however, $w \notin m_G(\emptyset)$. Notice that if $\emptyset$ is in everyone's neighborhood, then, according to the definition of C_G^2, every event will be common believed.

two versions of common belief. Heifetz (1996) proved that $\mathbf{EM_n^C}$ is *weakly complete*[9] with respect to the class of monotonic multi-agent neighborhood models. Lismont and Mongin (1994a, b) provide a weak completeness result for the second definition of common belief. Interestingly, Lismont and Mongin (2003) show that by weakening monotonicity, they can prove a *strong* completeness result for weak modal logics with the above common belief operators. The details of this fascinating result are beyond the scope of this section. Consult Heifetz (1999), Lismont (1995), and Barwise (1987) to clarify the relationship between the fixed-point and iterative definitions of common belief. Finally, van Benthem and Sarenac (2004) have an extensive discussion of different notions of group knowledge in the topological models from Sect. 1.4.1.

3.4 Dynamics with Neighborhoods: Game Logic

Suppose that two agents, say Abelard (A) and Eloise (E), are playing a game.[10] Say that Eloise can *force* a set of outcomes X if she has a strategy[11] such that, regardless of Abelard's choice of strategy, the outcome of the game will be some element of X.[12] This notion of forcing gives rise to a relation for each player between game states and sets of sets of outcomes. It is not hard to see that the collection of sets that a player can force is closed under supersets (if a player can force X and $X \subseteq Y$, then that player can force Y), but not closed under intersections (essentially, this follows from the fact that the *other* player can influence the outcome of the game).

Suppose that $W \neq \emptyset$ is a set of states, and let $\rho^A \subseteq \wp(W)$ and $\rho^E \subseteq \wp(W)$ be collections of sets that are intended to represent the powers of A and E, respectively, in some game. That is, each set in ρ^A is intended to be a set of outcomes that A can force by some strategy in a game. It is natural to impose the following constraints on these sets:

Monotonicity: For all sets of outcomes X and Y, for all players $i \in \{A, E\}$, if $X \in \rho^i$ and $X \subseteq Y$, then $Y \in \rho^i$.

---

[9]See Definition A.16 for the difference between strong and weak completeness. It is well known that, in general, logics with common belief operators are not compact, and so cannot be strongly complete.

[10]For this section, it is useful to have some basic understanding of game theory (Leyton-Brown and Shoham 2008). In Sect. 1.4.5, I introduced *strategic games* (Definition 1.39). In this section, I will also discuss *extensive games*. These are games in which the players do not necessarily make their choices simultaneously. I will restrict attention to two-player games in which the players take turns making moves. These are represented by *trees* in which the edges are labeled by actions, and each non-terminal node is associated with a player whose turn it is to move. The terminal nodes assign a payoff to each of the players.

[11]A **strategy** for Eloise (in an extensive game) is a function that assigns an action at each of her choice nodes.

[12]Recall the example used when discussing the logic of ability in Sect. 1.3.

3.4 Dynamics with Neighborhoods: Game Logic

Consistency: For any set of outcomes X, for each player $i \in \{E, A\}$, if i can force X, then the other player cannot force $W - X$. Formally, for all X and all $i \in \{E, A\}$, if $X \in \rho^i$, then $W - X \notin \rho^{-i}$.

In addition, in this section, I will restrict my analysis to games that are *determined* in the following sense:

Determined: For any set of outcomes X, for each player $i \in \{E, A\}$, if a player can force X, then the other player can force $W - X$. Formally, for all X and all $i \in \{E, A\}$, if $X \notin \rho^i$, then $W - X \in \rho^{-i}$, where $-i$ is the other player (i.e., if $i = E$, then $-i = A$).

For example, suppose that $W = \{1, 2, 3\}$, and consider the following two collections of subsets $\rho^A \subseteq \wp(W)$ and $\rho^E \subseteq \wp(W)$.

$\rho^E = \{\{1\}, \{2, 3\}, \{1, 2\}, \{1, 3\}, \{1, 2, 3\}\}$
$\rho^A = \{\{1, 2\}, \{1, 3\}, \{1, 2, 3\}\}$

As the reader is invited to check, these sets satisfy the above three properties. Van Benthem (2003) proved that any pairs of collections of sets satisfying the above three properties are the powers of the two players in some two-move finite game.[13] The main idea of the proof is to construct a game in which the actions for player i are the sets in the non-monotonic core of ρ^i. The outcome, given the choices for each player, is the intersection of the sets that each player chooses (the above properties guarantee that this set is non-empty). For instance, the above sets ρ^E and ρ^A are the sets that the players can force in the following game:

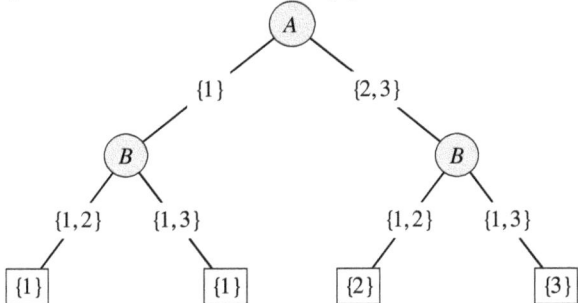

Exercise 3.16 Find at least two other games such that ρ^E and ρ^A are the sets that E and A can force. What can you say about the relationship between these different games?

Parikh (1985) introduced Game Logic as a generalization of *propositional dynamic logic* (PDL)[14] for reasoning about two-person determined games. In PDL,

---

[13] Cf. Bonanno (1992) for a related discussion.

[14] See Harel et al. (2000) for a discussion of propositional dynamic logic and its applications for reasoning about programs. See also Peleg (1987), Goldblatt (1992b), van Benthem et al. (2008) and Pacuit and Simon (2011) for variations of propositional dynamic logic related to topics discussed in this section.

the modalities are labeled with expressions that are intended to describe a program. That is, PDL includes formulas of the form $[\alpha]\varphi$, where α is a description of a program. The intended meaning of $\langle\alpha\rangle\varphi$ is that some execution of α ending in a state that satisfies φ is true, and the intended meaning of $[\alpha]\varphi$ is that every execution of α ends in a state satisfying φ. Parikh's main idea is to view the labels of the modalities as descriptions of two-person determined games. The modalities then describe the players' powers in the game. That is, $\langle\gamma\rangle\varphi$ is intended to mean that Eloise has a strategy in the game, described by γ, that guarantees that φ is true; and $[\gamma]\varphi$ is intended to mean that Abelard has a strategy in the game, described by γ, that guarantees that φ is true. In the remainder of this section, I will introduce game logic. Consult van Benthem (2014) and Pauly and Parikh (2003) for a more in-depth discussion of this logic and related topics.

Syntax and Semantics

Suppose that $\mathsf{Gm} = \{g_1, \ldots, g_n, \ldots\}$ is a finite or countable set of **primitive game expressions**, and At is a finite or countable set of atomic propositions. Formulas of game logic are generated by mutual recursion:

$$p \mid \neg\varphi \mid (\varphi \wedge \psi) \mid \langle\gamma\rangle\varphi$$

$$g \mid (\gamma; \gamma') \mid (\gamma \cup \gamma') \mid \gamma^* \mid \gamma^d$$

where $g \in \mathsf{Gm}$ and $p \in \mathsf{At}$. Let $\mathcal{L}^G$ denote the set of formulas of game logic. The Boolean connectives are defined as usual. Furthermore, define $[\gamma]\varphi$ ad $\neg\langle\gamma\rangle\neg\varphi$. Each game expression γ is intended to describe a determined game between two players. The intended meanings of the game operations are:

- $\gamma_1; \gamma_2$: First play γ_1; then play γ_2 (*game composition*).
- $\gamma_1 \cup \gamma_2$: Eloise chooses which of γ_1 or γ_2 to play (*choice*).
- γ^*: Eloise can choose how often to play γ (possibly not at all), and each time she has played γ, she can decide whether or not to play it again (*iteration*).
- $?\varphi$: Test whether φ currently holds (*test*).
- γ^d: Eloise and Abelard switch roles, then play γ (*dual*).

Using the dual operator, there are analogues of the choice and iteration operations for Abelard:

- $\gamma_1 \cap \gamma_2 := (\gamma_1^d \cup \gamma_2^d)^d$: Abelard chooses which of γ_1 or γ_2 to play.
- $\gamma^\times := ((\gamma^d)^*)^d$: Abelard can choose how often to play γ (possibly not at all), and each time he has played γ, he can decide whether or not to play it again.

The formula $\langle\gamma\rangle\varphi$ expresses that Eloise has a strategy in the game γ to achieve φ. Then, $[\gamma]\varphi$ means that Eloise does not have a strategy to guarantee that γ ends in a state satisfying $\neg\varphi$. By determinacy, this means that Abelard has a strategy to guarantee that the game γ ends in a state satisfying φ.

Models for the game logic language $\mathcal{L}^G$ include neighborhood functions for each primitive game operation. That is, a game logic model is a tuple $\langle W, \{N_g\}_{g \in \mathsf{Gm}}, V\rangle$,

3.4 Dynamics with Neighborhoods: Game Logic

where W is a non-empty set of states; for each $g \in \mathsf{Gm}$, $N_g : W \to \wp(\wp(W))$ is a monotonic neighborhood function (i.e., for all $w \in W$ and $X, Y \subseteq W$, if $X \in N_g(w)$ and $X \subseteq Y$, then $Y \in N_g(w)$); and $V : \mathsf{At} \to \wp(W)$ is a valuation function. It is convenient to treat the neighborhood functions as relations $N_g \subseteq W \times \wp(W)$ (cf. Remark 1.10).

Truth of the formulas from $\mathcal{L}^G$ at a state w in a model $\mathcal{M} = \langle W, \{N_g\}_{g \in \mathsf{Gm}}, V \rangle$ is defined as usual (cf. Definition 1.12). I give only the clause for the modal operator:

$$\mathcal{M}, w \models \langle \gamma \rangle \varphi \text{ iff } w N_\gamma [\![\varphi]\!]_\mathcal{M}.$$

where $N_\gamma \subseteq W \times \wp(W)$ is the relation associated with the complex game γ. This relation is defined by induction on the structure of γ. For any $N \subseteq W \times \wp(W)$, let $N^\leftarrow : \wp(W) \to \wp(W)$ denote the function where, for all $X \in \wp(W)$, $N^\leftarrow(X) = \{w \mid w N X\}$. To simplify the notation, I write N_γ instead of $N_\gamma^\leftarrow$. Given a set $\{N_g\}_{g \in \mathsf{Gm}}$ of neighborhood functions for the primitive game expressions, define relations N_γ for each game expression γ by recursion as follows:

$$\begin{aligned} N_{\gamma_1;\gamma_2}(Y) &= N_{\gamma_1}(N_{\gamma_2}(Y)) \\ N_{\gamma_1 \cup \gamma_2}(Y) &= N_{\gamma_1}(Y) \cup N_{\gamma_2}(Y) \\ N_{?\varphi}(Y) &= [\![\varphi]\!]_\mathcal{M} \cap Y \\ N_{\gamma^d}(Y) &= W - N_\gamma(W - Y) \\ N_{\gamma^*}(Y) &= \mu X. Y \cup N_\gamma(X) \end{aligned}$$

The last clause requires some additional explanation. Given a set $Y \subseteq W$ and a game expression γ, let $F_Y : \wp(X) \to \wp(X)$ be the function $F_{\gamma, Y}(X) = Y \cup N_\gamma(X)$. Then, $\mu X. Y \cup N_\gamma(X)$ is the **least fixed-point** of the function $F_{\gamma, Y}$.

Fixed Points

Consider any monotonic function on the nonempty set of states S—i.e., a function $F : \wp(S) \to \wp(S)$ such that for $X, Y \in \wp(S)$, $X \subseteq Y$ implies $F(X) \subseteq F(Y)$. A set $Z \subseteq S$ is a **fixed point** of F iff $F(Z) = Z$. Furthermore, Z is the **least (greatest) fixed-point** of F iff 1. Z is a fixed-point of F and 2. if Z' is a fixed-point of F, then $Z \subseteq Z'$ ($Z' \subseteq Z$). Let $\mu X. F(X)$ denote the least fixed-point of F and $\nu X. F(X)$ denote the greatest fixed-point of F. The famous Knaster–Tarski fixed-point theorem shows that fixed points of monotonic functions always exist:

Theorem 3.25 (Knaster–Tarski Fixed-Point Theorem) *Suppose that F is a monotonic function $F : \wp(S) \to \wp(S)$. Then, F has a least and greatest fixed-point:*

$$\begin{aligned} \mu X. F(X) &= \bigcap \{Y \mid Y \subseteq S, F(Y) = Y\} = \bigcap \{Y \mid Y \subseteq S, F(Y) \subseteq Y\} \\ \nu X. F(X) &= \bigcap \{Y \mid Y \subseteq S, F(Y) = Y\} = \bigcap \{Y \mid Y \subseteq S, F(Y) \supseteq Y\} \end{aligned}$$

Thus, if we can show that $N_\gamma : \wp(W) \to \wp(W)$ is monotonic for all γ, then by the Knaster–Tarski Fixed-Point Theorem, N_{γ^*} is guaranteed to exist. For the atomic game expressions $g \in \mathsf{Gm}$, it is assumed that N_g is monotonic. The proof that monotonicity lifts to all game expressions is left as an exercise.

Exercise 3.17 Prove that for all game expressions γ, $N_\gamma : \wp(W) \to \wp(W)$ is monotonic.

Results

Much of the work on game logic has focused on comparing the expressivity of the language to related logical systems, such as propositional dynamic logic (Pauly 2001; Parikh 1985) and the modal mu-calculus (Berwanger 2003; Pauly 2001; Parikh 1985). To get a sense of the expressive power of game logic, consider the following result from Parikh (1985). First of all, note that there are two ways for a player to satisfy a goal φ. The first way is to have a strategy that guarantees that φ is true. The second way is to find a strategy that forces the other player to violate the rules of the game. That is, $\langle \gamma \rangle \varphi$ is true if either E has a strategy in γ that forces φ to be true or if E has a strategy that forces the game into a position in which it is not possible for A to choose a move that conforms to the rules of the game described by γ. For example, suppose that a describes a game in which E is the only player to move, but E must make a move. Then, we can think of a as describing a relation R_a. Given this interpretation of a, the formula $\langle a \rangle \top$ is true at a state w as long as E has an available move at w (i.e., if there is an R_a-accessible state from w). Furthermore, the only way that $\langle a^d \rangle \bot$ is true at a state w is if there is no available move for A at the state w (i.e., w is a dead-end state for the relation R_a). Now, consider the formula $\langle (a^d)^* \rangle \bot$. The only way for this formula to be true is for E to keep playing the game a^d until A does not have an available move. Thus, $\langle (a^d)^* \rangle \bot$ is true iff the relation associated with a is well-founded (i.e., every set of states X has an element $w \in X$ such that $w\, R\, x$ for all $x \in X$). This is not something that can be expressed in PDL (though it can be expressed in the modal μ-calculus).

The main tools needed for a comprehensive study of the expressivity of game logic are the model constructions from Sect. 2.1. In particular, the appropriate notion of equivalence between game models is a monotonic bisimulation (Definition 2.2). It is straightforward to adapt this definition to models of game logic. In addition, it can be shown that the operations on game expressions (sequential composition, choice, iteration, and test) are *safe* for bisimulation. That is, the definition of a monotonic bisimulation for each primitive game expression $g \in \mathsf{Gm}$ can be lifted to all game expressions γ. The proof of this is left as an exercise.

Exercise 3.18 Suppose that $\mathcal{M} = \langle W, \{N_g\}_{g \in \mathsf{Gm}}, V \rangle$ and $\mathcal{M}' = \langle W', \{N'_g\}_{g \in \mathsf{Gm}}, V' \rangle$ are two models of game logic. Suppose that Z is a monotonic bisimulation between $\mathcal{M}$ and $\mathcal{M}'$. Prove that for all game expressions γ, Z is a monotonic bisimulation for N_γ. That is, show that for all game expressions γ, the neighborhood function N_γ satisfies the clauses in Definition 2.2.

I conclude this introduction to Game Logic with a brief discussion of axiomatizations. It is not hard to see that each modality $\langle \gamma \rangle$ is a monotonic modal operator. That is, for each game expression γ, the monotonicity rule is valid:

$$(\mathrm{Mon}_\gamma) \quad \frac{\varphi \to \psi}{\langle \gamma \rangle \varphi \to \langle \gamma \rangle \psi}$$

3.4 Dynamics with Neighborhoods: Game Logic

In addition, the following "reduction axioms" correspond to the construction rules for complex game expressions.

(composition) $\langle \gamma_1; \gamma_2 \rangle \varphi \leftrightarrow \langle \gamma_1 \rangle \langle \gamma_2 \rangle \varphi$
(choice) $\langle \gamma_1 \cup \gamma_2 \rangle \varphi \leftrightarrow \langle \gamma_1 \rangle \varphi \vee \langle \gamma_2 \rangle \varphi$
(fixedpoint) $\langle \gamma^* \rangle \varphi \leftrightarrow \varphi \vee \langle \gamma \rangle \langle \gamma^* \rangle \varphi$
(test) $\langle ?\psi \rangle \varphi \leftrightarrow \psi \wedge \varphi$
(dual) $\langle \gamma^d \rangle \leftrightarrow [\gamma]\varphi$

Finally, the following rule is valid since γ^* is interpreted as a *least* fixed-point operator:

$$\text{(LFP)} \quad \frac{\langle \gamma \rangle \varphi \to \varphi}{\langle \gamma^* \rangle \varphi \to \varphi}$$

Exercise 3.19 Prove that each of the above reduction axioms is valid over the class of game logic models. Also prove that the (LFP) rule is valid.

Let **GmL** be the smallest set of formulas that contains all tautologies in the language $\mathcal{L}^G$; is closed under Modus Ponens; is closed under the (Mon$_\gamma$) rules; contains all instances of the above reduction axiom schemes; and is closed under the (LFP) rule. Write **GmL**$^{-d}$ for **GmL** without the dual axiom (dual) and **GmL**$^{-*}$ for **GmL** without the fixed-point axiom (fixedpoint) and the (LFP) rule. The two main theorems about completeness of Game Logic are:

Theorem 3.26 (Parikh 1985) *For the language of game logic without the dual operator ($\cdot^d$), the logic **GmL**$^{-d}$ is sound and weakly complete with respect to the class of game models.*

Theorem 3.27 (Pauly 2001) *For the language of game logic without the fixed-point operator ($\cdot^*$), the logic **GmL**$^{-*}$ is sound and strongly complete with respect to the class of game models.*

The completeness of full game logic **GmL** has been open since Parikh introduced Game Logic in 1985.

Remark 3.28 (*The induction axiom*) Let **GmL$_K$** be the the logic that extends **GmL** with the additional axiom schemes:

$$\langle g \rangle(\varphi \vee \psi) \to (\langle g \rangle \varphi \vee \langle g \rangle \psi) \qquad \neg \langle g \rangle \bot$$

where $g \in \mathsf{Gm}$ is an atomic game expression. Familiar from the literature on propositional dynamic logic (Harel et al. 2000), the induction axiom is:

$$\langle \gamma^* \rangle \varphi \to (\varphi \vee \langle \gamma^* \rangle(\neg \varphi \wedge \langle \gamma \rangle \varphi))$$

See Pauly (2001, Sect. 7.3) for a proof that the induction axiom is derivable in **GmL$_K^{-d}$**, but not in **GmL$_K$**.

3.5 Dynamics on Neighborhood Structures

The final section of this book is focused on recent dynamic logics of knowledge update (van Benthem 2011) and belief revision (van Benthem 2004; Baltag and Smets 2006b).[15] These logics include informational actions that *change* the models. Examples range from "hard" information provided by *public announcements* or public observations (Plaza 1989; Gerbrandy 1999) to softer signals encoding different policies of belief revision (cf. Rott 2006) by radical or conservative *upgrades* of plausibility orderings. Many of the ideas from this literature can be adapted to the neighborhood setting (cf. Zvesper 2010; van Ditmarsch et al. 2015; van Benthem and Pacuit 2011; Ma and Sano 2015). A complete overview of this literature is beyond the scope of this book. In this section, I focus on two key topics from this research area. The first subsection discusses the definition of public announcements in neighborhood models. The second subsection is an abbreviated discussion of evidence dynamics from van Benthem and Pacuit (2011).

3.5.1 Public Announcements

The simplest type of information change is receiving information from an infallible source. This operation, called *public announcement* (Plaza 1989), transforms a model by removing all states where the announced formula is false. Following the presentation in Ma and Sano (2015), I explain two different ways to define public announcements in neighborhood models.

For simplicity, in this section, I focus on single-agent neighborhood models. Suppose that $\mathcal{M} = \langle W, N, V \rangle$ is a monotonic neighborhood model. A public announcement of φ in $\mathcal{M}$ transforms $\mathcal{M}$ into a submodel in which the set of states is $[\![\varphi]\!]_{\mathcal{M}}$. The question is how to define the neighborhood function in this submodel. Ma and Kato (2015) propose two different ways to answer this question.

Definition 3.29 (*Intersection/Subset Submodel*) Let $\mathcal{M} = \langle W, N, V \rangle$ be a monotonic neighborhood model (with At the set of atomic propositions) and $\emptyset \neq X \subseteq W$. The **intersection (subset) submodel of** $\mathcal{M}$ is $\mathcal{M}^{\star X} = \langle X, N^{\star X}, V^{\star X} \rangle$, for $\star \in \{\cap, \subseteq\}$, where for all $p \in \text{At}$, $V^{\star X}(p) = V(p) \cap X$ and $N^{\star X} : X \to \wp(\wp(X))$ are defined as follows: for all $w \in X$,

$$N^{\cap X}(w) = \{Y \mid Y = Z \cap X \text{ for some } Z \in N(w)\}.$$

$$N^{\subseteq X}(w) = \{Y \mid Y \subseteq X \text{ and } Y \in N(w)\}.$$

We write $\mathcal{M}^{\star \varphi}$ for $\mathcal{M}^{\star [\![\varphi]\!]_{\mathcal{M}}}$.

---

[15]Consult Pacuit (2013b) for a shorter survey of this research area.

3.5 Dynamics on Neighborhood Structures

Exercise 3.20 Suppose that $\mathcal{F} = \langle W, N \rangle$ is monotonic. Prove that $\mathcal{F}^{\cap X} = \langle X, N^{\cap X} \rangle$ and $\mathcal{F}^{\subseteq X} = \langle X, N^{\subseteq X} \rangle$ are both monotonic.

Exercise 3.21 Show that the following properties are preserved by the $(.)^{\cap X}$ and $(.)^{\subseteq X}$ operations on models (cf. Ma and Sano (2015), Proposition 3).

- For each $w \in W$ and $X, Y \subseteq W$, if $X \in N(w)$ and $Y \in N(w)$, then $X \cap Y \in N(w)$.
- For each $w \in W$, $w \in \bigcap N(w)$.
- For each $w \in W$, if $X \in N(w)$, then $\{v \mid X \in N(v)\} \in N(w)$.

To reason about the above model transformations, we extend the basic modal language with public announcement operators. Suppose that At is a set of atomic propositions. The full language contains modal operators corresponding to the two operations from Definition 3.29:

$$p \mid \neg \varphi \mid (\varphi \wedge \psi) \mid \Box \varphi \mid [\varphi]^{\cap} \psi \mid [\varphi]^{\subseteq} \psi$$

where $p \in$ At. The additional Boolean connectives are defined as usual. Let $\mathcal{L}^{\cap}$ be the fragment of the above language in which the only dynamic modalities are $[\varphi]^{\cap}$, and $\mathcal{L}^{\subseteq}$ is the fragment in which the only dynamic modalities are $[\varphi]^{\subseteq}$. Truth is defined as in Definition 1.12, with the following clauses for the new modalities:

- $\mathcal{M}, w \models [\varphi]^{\cap} \psi$ iff $\mathcal{M}, w \models \varphi$ implies that $\mathcal{M}^{\cap \varphi}, w \models \psi$.
- $\mathcal{M}, w \models [\varphi]^{\subseteq} \psi$ iff $\mathcal{M}, w \models \varphi$ implies that $\mathcal{M}^{\subseteq \varphi}, w \models \psi$.

For $\star \in \{\cap, \subseteq\}$, let $\langle \varphi \rangle^{\star} \psi$ be $\neg [\varphi]^{\star} \neg \psi$. The truth clause for this formula is:

$$\mathcal{M}, w \models \langle \varphi \rangle^{\star} \psi \text{ iff } \mathcal{M}, w \models \varphi \text{ and } \mathcal{M}^{\star \varphi}, w \models \psi.$$

The standard approach to axiomatizing languages with public announcement operators is to find so-called *recursion axioms*. Recursion axioms provide an insightful syntactic analysis of public announcements that complements the semantic analysis: The recursion axioms describe the effect of an announcement in terms of what is true before the announcement. Consult van Benthem (2011) for a general discussion of the recursion-axiom methodology. The recursion axioms for the languages $\mathcal{L}^{\cap}$ and $\mathcal{L}^{\subseteq}$ are:

$$
\begin{aligned}
&\text{(PA1)} && [\varphi]^{\star} p && \leftrightarrow (\varphi \to p) && (p \in \text{At}, \star \in \{\cap, \subseteq\}) \\
&\text{(PA2)} && [\varphi]^{\star}(\psi \wedge \chi) && \leftrightarrow ([\varphi]^{\star}\psi \wedge [\varphi]^{\star}\chi) && (\star \in \{\cap, \subseteq\}) \\
&\text{(PA3)} && [\varphi]^{\star} \neg \psi && \leftrightarrow (\varphi \to \neg [\varphi]^{\star} \psi) && (\star \in \{\cap, \subseteq\}) \\
&\text{(PA4)} && [\varphi]^{\star} [\psi]^{\star} \chi && \leftrightarrow [\varphi \wedge [\varphi]^{\star} \psi] \chi && (\star \in \{\cap, \subseteq\}) \\
&\text{(PA}\cap) && [\varphi]^{\cap} \Box \psi && \leftrightarrow (\varphi \to \Box [\varphi]^{\cap} \psi) \\
&\text{(PA}\subseteq) && [\varphi]^{\subseteq} \Box \psi && \leftrightarrow (\varphi \to \Box \langle \varphi \rangle^{\subseteq} \psi)
\end{aligned}
$$

Lemma 3.7 *The formula* $[\varphi]^{\cap} \Box \psi \leftrightarrow (\varphi \to \Box [\varphi]^{\cap} \psi)$ *is valid on monotonic models.*

Proof Suppose that $\mathcal{M} = \langle W, N, V \rangle$ is a monotonic neighborhood model. First, note that an immediate consequence of the truth of $[\varphi]^\cap \psi$ is that:

$$(*) \quad [\![[\varphi]^\cap \psi]\!]_\mathcal{M} = (W - [\![\varphi]\!]_\mathcal{M}) \cup [\![\psi]\!]_{\mathcal{M}^{\cap\varphi}}.$$

Claim 1: $\mathcal{M}, w \models [\varphi]^\cap \Box \psi \to (\varphi \to \Box[\varphi]^\cap \psi)$. *Proof*: Suppose that $\mathcal{M}, w \models [\varphi]^\cap \Box \psi$ and $\mathcal{M}, w \models \varphi$. Then, $\mathcal{M}^{\cap\varphi}, w \models \Box \psi$. This implies that $[\![\psi]\!]_{\mathcal{M}^{\cap\varphi}} \in N^{\cap\varphi}(w)$. By the definition of $N^{\cap\varphi}$, there is a $Y \in N(w)$ such that $[\![\psi]\!]_{\mathcal{M}^{\cap\varphi}} = Y \cap [\![\varphi]\!]_\mathcal{M}$. By basic set theory, we have that $Y \subseteq (W - [\![\varphi]\!]_\mathcal{M}) \cup [\![\psi]\!]_{\mathcal{M}^{\cap\varphi}}$. Since N is monotonic and $Y \in N(w)$, we have that $(W - [\![\varphi]\!]_\mathcal{M}) \cup [\![\psi]\!]_{\mathcal{M}^{\cap\varphi}} \in N_i(w)$. Therefore, by $(*)$, $\mathcal{M}, w \models \Box[\varphi]^\cap \psi$. Therefore, $\mathcal{M}, w \models [\varphi]^\cap \Box \psi \to (\varphi \to \Box[\varphi]^\cap \psi)$.

Claim 2: $\mathcal{M}, w \models (\varphi \to \Box[\varphi]^\cap \psi) \to [\varphi]^\cap \Box \psi$. *Proof*: Suppose that $\mathcal{M}, w \models \varphi \to \Box[\varphi]^\cap \psi$. We must show that $\mathcal{M}, w \models [\varphi]^\cap \Box \psi$. That is, we must show that $\mathcal{M}^{\cap\varphi}, w \models \Box \psi$. Suppose that $\mathcal{M}, w \models \varphi$. Therefore, $\mathcal{M}, w \models \Box[\varphi]^\cap \psi$. This implies that $[\![[\varphi]^\cap \psi]\!]_\mathcal{M} \in N(w)$. By $(*)$ and the fact that $[\![\psi]\!]_{\mathcal{M}^{\cap\varphi}} \subseteq [\![\varphi]\!]_\mathcal{M}$, we have that

$$\begin{aligned}[][\![[\varphi]^\cap \psi]\!]_\mathcal{M} \cap [\![\varphi]\!]_\mathcal{M} &= ((W - [\![\varphi]\!]_\mathcal{M}) \cup [\![\psi]\!]_{\mathcal{M}^{\cap\varphi}}) \cap [\![\varphi]\!]_\mathcal{M} \\ &= ((W - [\![\varphi]\!]_\mathcal{M}) \cap [\![\varphi]\!]_\mathcal{M}) \cup ([\![\psi]\!]_{\mathcal{M}^{\cap\varphi}} \cap [\![\varphi]\!]_\mathcal{M}) \\ &= [\![\psi]\!]_{\mathcal{M}^{\cap\varphi}} \cap [\![\varphi]\!]_\mathcal{M} \\ &= [\![\psi]\!]_{\mathcal{M}^{\cap\varphi}}. \end{aligned}$$

Since $[\![[\varphi]^\cap \psi]\!]_\mathcal{M} \in N(w)$, we have that $[\![\psi]\!]_{\mathcal{M}^{\cap\varphi}} \in N^{\cap\varphi}(w)$. Thus, $\mathcal{M}^{\cap\varphi}, w \models \Box \psi$. Therefore, $\mathcal{M}, w \models [\varphi]^\cap \Box \psi$; and so, $\mathcal{M}, w \models (\varphi \to \Box[\varphi]^\cap \psi) \to [\varphi]^\cap \Box \psi$. □

Lemma 3.8 *The formula* $[\varphi]^\subseteq \Box \psi \leftrightarrow (\varphi \to \Box\langle\varphi\rangle^\subseteq \psi)$ *is valid on monotonic models.*

Proof Suppose that $\mathcal{M} = \langle W, N, V \rangle$ is a monotonic neighborhood model. First, note that, by construction of $\mathcal{M}^{\subseteq\varphi}$, we have that $[\![\psi]\!]_{\mathcal{M}^{\subseteq\varphi}} \subseteq [\![\varphi]\!]_\mathcal{M}$. Thus, $[\![\psi]\!]_{\mathcal{M}^{\subseteq\varphi}} \cap [\![\varphi]\!]_\mathcal{M} = [\![\psi]\!]_{\mathcal{M}^{\subseteq\varphi}}$. Then, by the definition of truth for $\langle\varphi\rangle^\subseteq \psi$,

$$[\![\langle\varphi\rangle^\subseteq \psi]\!]_\mathcal{M} = [\![\varphi]\!]_\mathcal{M} \cap [\![\psi]\!]_{\mathcal{M}^{\subseteq\varphi}} = [\![\psi]\!]_{\mathcal{M}^{\subseteq\varphi}}.$$

In addition, for all formulas φ and ψ, we have that

$$\begin{aligned}[][\![\varphi \wedge [\varphi]^\subseteq \psi]\!]_\mathcal{M} &= [\![\varphi]\!]_\mathcal{M} \cap [\![[\varphi]^\subseteq \psi]\!]_\mathcal{M} \\ &= [\![\varphi]\!]_\mathcal{M} \cap ((W - [\![\varphi]\!]_\mathcal{M}) \cup [\![\psi]\!]_{\mathcal{M}^{\subseteq\varphi}}) \\ &= ([\![\varphi]\!]_\mathcal{M} \cap (W - [\![\varphi]\!]_\mathcal{M})) \cup ([\![\varphi]\!]_\mathcal{M} \cap [\![\psi]\!]_{\mathcal{M}^{\subseteq\varphi}}) \\ &= \emptyset \cup ([\![\varphi]\!]_\mathcal{M} \cap [\![\psi]\!]_{\mathcal{M}^{\subseteq\varphi}}) \\ &= [\![\varphi]\!]_\mathcal{M} \cap [\![\psi]\!]_{\mathcal{M}^{\subseteq\varphi}} \\ &= [\![\psi]\!]_{\mathcal{M}^{\subseteq\varphi}}. \end{aligned}$$

Putting everything together, we have that

3.5 Dynamics on Neighborhood Structures

$$(*) \quad [\![\langle\varphi\rangle^{\subseteq}\psi]\!]_{\mathcal{M}} = [\![\varphi \wedge [\varphi]^{\subseteq}\psi]\!]_{\mathcal{M}} = [\![\psi]\!]_{\mathcal{M}^{\subseteq\varphi}}.$$

Claim 1: $\mathcal{M}, w \models [\varphi]^{\subseteq}\Box\psi \to (\varphi \to \Box\langle\varphi\rangle^{\subseteq}\psi)$ *Proof.* Suppose that $\mathcal{M}, w \models [\varphi]^{\subseteq}\Box\psi$ and $\mathcal{M}, w \models \varphi$. Then, $\mathcal{M}^{\subseteq\varphi}, w \models \Box\psi$. This means that $[\![\psi]\!]_{\mathcal{M}^{\subseteq\varphi}} \in N^{\subseteq\varphi}(w)$. This means that $[\![\psi]\!]_{\mathcal{M}^{\subseteq\varphi}} \subseteq [\![\varphi]\!]_{\mathcal{M}}$ and $[\![\psi]\!]_{\mathcal{M}^{\subseteq\varphi}} \in N(w)$. By $(*)$, $[\![\langle\varphi\rangle^{\subseteq}\psi]\!]_{\mathcal{M}} = [\![\psi]\!]_{\mathcal{M}^{\subseteq\varphi}} \in N(w)$. Hence, $\mathcal{M}, w \models \Box\langle\varphi\rangle^{\subseteq}\psi$, as desired.

Claim 2: $\mathcal{M}, w \models (\varphi \to \Box\langle\varphi\rangle^{\subseteq}\psi) \to [\varphi]^{\subseteq}\Box\psi$ *Proof.* Suppose that $\mathcal{M}, w \models \varphi \to \Box\langle\varphi\rangle^{\subseteq}\psi$. We must show that $\mathcal{M}, w \models [\varphi]^{\subseteq}\Box\psi$. Assume that $\mathcal{M}, w \models \varphi$. Then, $\mathcal{M}, w \models \Box\langle\varphi\rangle^{\subseteq}\psi$. Thus, $[\![\langle\varphi\rangle^{\subseteq}\psi]\!]_{\mathcal{M}} \in N(w)$. By $(*)$, $[\![\langle\varphi\rangle^{\subseteq}\psi]\!]_{\mathcal{M}} = [\![\varphi \wedge [\varphi]^{\subseteq}\psi]\!]_{\mathcal{M}} \subseteq [\![\varphi]\!]_{\mathcal{M}}$. By the definition of $N^{\subseteq\varphi}$ and $(*)$, we have that $[\![\psi]\!]_{\mathcal{M}^{\subseteq\varphi}} = [\![\langle\varphi\rangle^{\subseteq}\psi]\!]_{\mathcal{M}} \in N^{\subseteq\varphi}(w)$. Thus, $\mathcal{M}^{\subseteq\varphi}, w \models \Box\psi$. Hence, $\mathcal{M}, w \models [\varphi]^{\subseteq}\Box\psi$, as desired. $\square$

Exercise 3.22 Suppose that $\mathcal{M} = \langle W, N, V \rangle$ is augmented. What is the relationship between $\mathcal{M}^{\cap\varphi}$ and $\mathcal{M}^{\subseteq\varphi}$?

Consult Ma and Sano (2015) for more details about the above two versions of public announcements. I conclude this section by discussing another variant of the public announcement transformation. It is not hard to find a model $\mathcal{M} = \langle W, N, V \rangle$ and a formula such that $\emptyset \in N^{\cap\varphi}(w)$, but $\emptyset \notin N(w)$. In such a case, the public announcement of φ makes the agent believe $\bot$. An alternative approach ignores any inconsistencies with the announced formula.[16] Suppose that $\mathcal{M} = \langle W, N, V \rangle$ is a monotonic neighborhood model. Given a set $X \subseteq W$, let $\mathcal{M}^{\cap X} = \langle X, N^{\cap X}, V^{\cap X}\rangle$, where $V^{\cap X}$ is the restriction of V to X, and $N^{\cap X}$ is defined as follows, for each $w \in X$:

$$N^{\cap X}(w) = \{Y \mid \emptyset \neq Y = X \cap Z \text{ for some } Z \in N(w)\}.$$

Following the methodology described above, extend the basic modal language $\mathcal{L}$ with modalities $[\varphi]^{\cap}\psi$ with truth defined in the obvious way:

$$\mathcal{M}, w \models [\varphi]^{\cap}\psi \text{ iff if } \mathcal{M}, w \models \varphi, \text{ then } \mathcal{M}^{\cap\varphi}, w \models \psi.$$

Interestingly, the variant of $PA\cap$ is *not* valid.

Observation 3.30 *The formula* $[\varphi]^{\cap}\Box\psi \leftrightarrow (\varphi \to \Box[\varphi]^{\cap}\psi)$ *is not valid.*

Proof Suppose that $\mathcal{M} = \langle W, N, V \rangle$, where $W = \{w, v\}$, $V(p) = \{w\}$, and $N(w) = N(v) = \{\{w\}, \{v\}, \{w, v\}\}$. Then, $N^{\cap p}(w) = \{\{w\}\}$. Thus, $\mathcal{M}^{\cap p} = \langle\{w\}, N^{\cap p}, V\rangle$. Then, since $[\![[p]^{\cap}\bot]\!]_{\mathcal{M}} = [\![p \to \bot]\!]_{\mathcal{M}} = [\![\neg p]\!]_{\mathcal{M}} = \{v\} \in N(w)$, we have that $\mathcal{M}, w \models \Box[p]^{\cap}\bot$. However, since $\emptyset = [\![\bot]\!]_{\mathcal{M}^{\cap p}} \notin N^{\cap p}(w)$, we have that $\mathcal{M}^{\cap p}, w \not\models [p]^{\cap}\Box\bot$. Thus, $\mathcal{M}, w \not\models p \to [p]^{\cap}\Box\bot$, but $\mathcal{M}, w \models \Box[p]^{\cap}\bot$. $\square$

---

[16]The belief revision literature (Alchourrón et al. 1985) contains a number of different ways to change an agent's beliefs given an observation that is inconsistent with the agent's current beliefs (cf. van Benthem 2011).

Recursion axioms for this version of the public announcement operator requires a simplified version of the conditional modality from Sect. 1.4.3.

Definition 3.31 (*Conditional neighborhood modality*) For formulas φ and ψ, let $\square^\varphi \psi$ be a formula, called a **conditional neighborhood modality**, interpreted at states in a neighborhood model $\mathcal{M} = \langle W, N, V \rangle$ as follows:

- $\mathcal{M}, w \models \square^\varphi \psi$ iff there is a $X \in N(w)$ such that $X \cap [\![\varphi]\!]_\mathcal{M} \neq \emptyset$ and for all $v \in X \cap [\![\varphi]\!]_\mathcal{M}$, $\mathcal{M}, v \models \psi$.

Exercise 3.23 Prove that $\square^\varphi \psi$ is not equivalent to $\square(\varphi \to \psi)$.

In this extended language, we can state recursive axioms. Of course, we need recursive axioms for the new modality in addition to formulas of the form $\square \psi$.

$$(PA_{\mathfrak{m}1}) \quad [\varphi]^{\mathfrak{m}} \square \psi \leftrightarrow (\varphi \to \square^\varphi [\varphi]^{\mathfrak{m}} \psi)$$
$$(PA_{\mathfrak{m}2}) \quad [\varphi]^{\mathfrak{m}} \square^\alpha \psi \leftrightarrow (\varphi \to \square^{\varphi \wedge [\varphi]^{\mathfrak{m}} \alpha} [\varphi]^{\mathfrak{m}} \psi)$$

I will show that $PA_{\mathfrak{m}2}$ is valid and leave the verification that $PA_{\mathfrak{m}1}$ is valid as an exercise.

Lemma 3.9 *The formula* $[\varphi]^{\mathfrak{m}} \square^\alpha \psi \leftrightarrow (\varphi \to \square^{\varphi \wedge [\varphi]^{\mathfrak{m}} \alpha} [\varphi]^{\mathfrak{m}} \psi)$ *is valid on monotonic neighborhood models.*

Proof Suppose that $\mathcal{M} = \langle W, N, V \rangle$ is a monotonic neighborhood model.

Claim 1: $\mathcal{M}, w \models [\varphi]^{\mathfrak{m}} \square^\alpha \psi \to (\varphi \to \square^{\varphi \wedge [\varphi]^{\mathfrak{m}} \alpha} [\varphi]^{\mathfrak{m}} \psi)$. *Proof*: Suppose that $\mathcal{M}, w \models [\varphi]^{\mathfrak{m}} \square^\alpha \psi$ and that $\mathcal{M}, w \models \varphi$. Then, $\mathcal{M}^{\mathfrak{m}\varphi}, w \models \square^\alpha \psi$. This implies that there is a $X \in N^{\mathfrak{m}\varphi}(w)$ such that $X \cap [\![\alpha]\!]_{\mathcal{M}^{\mathfrak{m}\varphi}} \neq \emptyset$ and $X \cap [\![\alpha]\!]_{\mathcal{M}^{\mathfrak{m}\varphi}} \subseteq [\![\psi]\!]_{\mathcal{M}^{\mathfrak{m}\varphi}}$. Since $X \subseteq [\![\varphi]\!]_\mathcal{M}$ and $[\![[\varphi]^{\mathfrak{m}}\alpha]\!]_\mathcal{M} = (W - [\![\varphi]\!]_\mathcal{M}) \cup [\![\alpha]\!]_{\mathcal{M}^{\mathfrak{m}\varphi}}$, we have that $X \cap [\![[\varphi]^{\mathfrak{m}}\alpha]\!]_\mathcal{M} = X \cap [\![\alpha]\!]_{\mathcal{M}^{\mathfrak{m}}}$. Furthermore, since $X \in N^{\mathfrak{m}\varphi}(w)$, there is a $Y \in N(w)$ such that $\emptyset \neq X = Y \cap [\![\varphi]\!]_\mathcal{M}$. Hence,

$$Y \cap [\![\varphi \wedge [\varphi]^{\mathfrak{m}}\alpha]\!]_\mathcal{M} = Y \cap [\![\varphi]\!]_\mathcal{M} \cap [\![[\varphi]^{\mathfrak{m}}\alpha]\!]_\mathcal{M}$$
$$= X \cap [\![[\varphi]^{\mathfrak{m}}\alpha]\!]_\mathcal{M}$$
$$= X \cap [\![\alpha]\!]_{\mathcal{M}^{\mathfrak{m}\varphi}}$$
$$\subseteq [\![\psi]\!]_{\mathcal{M}^{\mathfrak{m}\varphi}}$$

Therefore, since $[\![\psi]\!]_{\mathcal{M}^{\mathfrak{m}\varphi}} \subseteq [\![[\varphi]^{\mathfrak{m}}\psi]\!]_\mathcal{M}$ and $X \cap [\![\varphi]\!]_\mathcal{M} \cap [\![[\varphi]^{\mathfrak{m}}\alpha]\!]_\mathcal{M} \neq \emptyset$, it follows that $\mathcal{M}, w \models \square^{\varphi \wedge [\varphi]^{\mathfrak{m}}\alpha} \psi$.

Claim 2: $\mathcal{M}, w \models (\varphi \to \square^{\varphi \wedge [\varphi]^{\mathfrak{m}}\alpha} [\varphi]^{\mathfrak{m}} \psi) \to [\varphi]^{\mathfrak{m}} \square^\alpha \psi$. *Proof*: Suppose that $\mathcal{M}, w \models \varphi \to \square^{\varphi \wedge [\varphi]^{\mathfrak{m}}\alpha} [\varphi]^{\mathfrak{m}} \psi$. We must show that $\mathcal{M}, w \models [\varphi]^{\mathfrak{m}} \square^\alpha \psi$. Suppose that $\mathcal{M}, w \models \varphi$. Then, $\mathcal{M}, w \models \square^{\varphi \wedge [\varphi]^{\mathfrak{m}}\alpha} [\varphi]^{\mathfrak{m}} \psi$. Then, there is a $X \in N(w)$ such that $X \cap [\![\varphi \wedge [\varphi]^{\mathfrak{m}}\alpha]\!]_\mathcal{M} \neq \emptyset$ and $X \cap [\![\varphi \wedge [\varphi]^{\mathfrak{m}}\alpha]\!]_\mathcal{M} \subseteq [\![[\varphi]^{\mathfrak{m}}\psi]\!]_\mathcal{M}$. Then, $Y = X \cap [\![\varphi]\!]_\mathcal{M} \in N^{\mathfrak{m}\varphi}(w)$. Since $[\![[\varphi]^{\mathfrak{m}}\chi]\!]_\mathcal{M} = (W - [\![\varphi]\!]_\mathcal{M}) \cup [\![\chi]\!]_{\mathcal{M}^{\mathfrak{m}\varphi}}$ (for $\chi = \alpha, \psi$), we have

$$\emptyset \neq Y \cap [\![\alpha]\!]_{\mathcal{M}^{\mathfrak{m}\varphi}} \subseteq [\![\psi]\!]_{\mathcal{M}^{\mathfrak{m}\varphi}}$$

Thus, $\mathcal{M}^{\mathfrak{m}\varphi}, w \models \square^\alpha \psi$, and so $\mathcal{M}, w \models [\varphi]^{\mathfrak{m}} \square^\alpha \psi$. $\square$

3.5.2 Evidence Dynamics

The evidence models from Sect. 1.4.4 are a natural setting in which to further study dynamic operations on neighborhood models. In this final section of the book, I introduce the dynamic logics of "evidence management" from van Benthem and Pacuit (2011). The reader is invited to review the definition of evidence models (Definition 1.35) and truth for the language $\mathcal{L}^{ev}$ (Definition 1.37).

From the perspective of evidence models, the simple public announcement operation introduced in the previous section is actually a compound of various transformations. A public announcement of φ can be naturally "deconstructed" into a complex combination of three distinct operations:

1. **Evidence addition**: the agent accepts that φ is an "admissible" piece of evidence (perhaps on par with the other available evidence).
2. **Evidence removal**: the agent removes any evidence for $\neg\varphi$.
3. **Evidence modification**: the agent incorporates φ into each piece of evidence gathered so far, making φ the most important piece of evidence.

The evidence models from Sect. 1.4.4 allow us to study each of these operations individually. In the remainder of this section, I briefly discuss the first two operations (see van Benthem and Pacuit 2011, for a complete discussion).

Let $\mathcal{M}$ be an evidence model and φ a new piece of evidence that the agent decides to *accept*. Here, "acceptance" does not necessarily mean that the agent believes that φ is true, but, rather, that she agrees that φ should be considered when "weighing" her evidence. The formal definition of this action is straightforward:

Definition 3.32 (*Evidence Addition*) Suppose that $\mathcal{M} = \langle W, E, V \rangle$ is an evidence model, and φ a formula in $\mathcal{L}^{ev}$. The model $\mathcal{M}^{+\varphi} = \langle W^{+\varphi}, E^{+\varphi}, V^{+\varphi} \rangle$ has $W^{+\varphi} = W$, $V^{+\varphi} = V$ and for all $w \in W$,

$$E^{+\varphi}(w) = E(w) \cup \{[\![\varphi]\!]_{\mathcal{M}}\}.$$

This operation can be described explicitly with a dynamic modality $[+\varphi]\psi$, which is intended to mean that "ψ is true after φ is accepted as an admissible piece of evidence". The truth condition for this formula is straightforward:

(EA) $\quad \mathcal{M}, w \models [+\varphi]\psi \;\;\text{iff}\;\; [\![\varphi]\!]_{\mathcal{M}} \neq \emptyset \text{ implies } \mathcal{M}^{+\varphi}, w \models \psi.$

Since evidence sets cannot be empty, the precondition for evidence addition is that φ is true at some state. Compare this with the precondition for public announcement from the previous section, which requires the accepted formula to be true. The recursion axioms for languages containing only the evidence and universal modalities are:

$$[+\varphi]p \leftrightarrow (\langle A\rangle\varphi \to p) \quad (p \in \mathsf{At})$$
$$[+\varphi](\psi \wedge \chi) \leftrightarrow ([+\varphi]\psi \wedge [+\varphi]\chi)$$
$$[+\varphi]\neg\psi \leftrightarrow (\langle A\rangle\varphi \to \neg[+\varphi]\psi)$$
$$[+\varphi]\langle\psi \leftrightarrow (\langle A\rangle\varphi \to (\langle[+\varphi]\psi \vee [A](\varphi \to [+\varphi]\psi)))$$
$$[+\varphi][A]\psi \leftrightarrow (\langle A\rangle\varphi \to [A][+\varphi]\psi)$$

Exercise 3.24 Prove that the above axioms are valid on any evidence model.

The complete logical analysis of even this simple operation is surprisingly subtle. Finding a similar recursion law for the belief operator in $\mathcal{L}^{ev}$ requires an extension to $\mathcal{L}^{ev}$. See van Benthem and Pacuit (2011) for a discussion.

With a public announcement of φ, the agent also agrees to *ignore* states inconsistent with φ. The latter attitude suggests an act of *evidence removal* as a natural converse to addition. While "removal" has been a challenge to dynamic-epistemic logics, our richer setting suggests a natural logic.

Definition 3.33 (*Evidence Removal*) Let $\mathcal{M} = \langle W, E, V\rangle$ be an evidence model, and φ a formula in $\mathcal{L}^{ev}$. The model $\mathcal{M}^{-\varphi} = \langle W^{-\varphi}, E^{-\varphi}, V^{-\varphi}\rangle$ has $W^{-\varphi} = W$, $V^{-\varphi} = V$, and for all $w \in W$,

$$E^{-\varphi}(w) = E(w) - \{X \mid X \subseteq [\![\varphi]\!]_\mathcal{M}\}.$$

This time, the corresponding dynamic modality is $[-\varphi]\psi$ ("after removing the evidence that φ, ψ is true"), defined as follows[17]:

(ER) $\quad \mathcal{M}, w \models [-\varphi]\psi$ iff $[\![\varphi]\!]_\mathcal{M} \neq W$ implies $\mathcal{M}^{-\varphi}, w \models \psi$

Finding a recursion axiom for this dynamic modality requires an extension to the modal language $\mathcal{L}^{ev}$. To motivate the language extension, consider the formula $[-\varphi]\langle\ ]\psi$. This formula is true at a state w when there is evidence at w for ψ that survived the removal of φ. Now, the evidence that survives the removal of φ is the evidence that contains states in which φ is false. This suggests a new conditional modality: $\square(\varphi; \psi)$ that is true at a state w in an evidence model $\mathcal{M} = \langle W, N, V\rangle$ provided that there is an $X \in E(w)$ such that $X \cap [\![\varphi]\!]_\mathcal{M} \neq \emptyset$ and $X \subseteq [\![\psi]\!]_\mathcal{M}$. More generally,

Definition 3.34 (*Instantial neighborhood modality*) Suppose that $\varphi_1, \ldots, \varphi_k, \psi$ are modal formulas (e.g., formulas from $\mathcal{L}$ or $\mathcal{L}^{ev}$). Let $\square(\varphi_1, \ldots, \varphi_k; \psi)$ be a formula (called an instantial neighborhood modality) interpreted in a neighborhood model $\mathcal{M} = \langle W, N, V\rangle$ as follows:

---

[17] Removing the evidence for φ is weaker than the usual notion of *contracting* one's beliefs by φ in the theory of belief revision (Rott 2001). It is possible to remove the evidence for φ and yet the agent maintains her belief in φ. Formally, $[-\varphi]\neg B\varphi$ is not valid. To see this, let $W = \{w_1, w_2, w_3\}$ with p true only at w_3. Consider an evidence model with two pieces of evidence: $\mathcal{E} = \{\{w_1, w_3\}, \{w_2, w_3\}\}$. The agent believes p and, since the model does not change when removing the evidence for p, $[-p]Bp$ is true. The same is true for the model with explicit evidence for p—i.e., $\mathcal{E}' = \{\{w_1, w_3\}, \{w_2, w_3\}, \{w_3\}\}$.

3.5 Dynamics on Neighborhood Structures

$\mathcal{M}, w \models \Box(\varphi_1, \ldots, \varphi_k; \psi)$ iff there is an $X \in N(w)$ such that $X \subseteq [\![\psi]\!]_{\mathcal{M}}$ and for all $i = 1, \ldots, k$, $X \cap [\![\varphi_i]\!]_{\mathcal{M}} \neq \emptyset$.

Exercise 3.25 Prove that $\Box(\varphi_1, \ldots, \varphi_k; \psi) \rightarrow \Box(\varphi_1, \ldots, \varphi_k, \alpha; \psi) \vee \Box(\varphi_1, \ldots, \varphi_k; \psi \wedge \neg\alpha)$ is valid on any neighborhood model (see van Benthem et al. 2017, Axiom NT from Sect. 4).

It turns out that the $\Box(\varphi_1, \ldots, \varphi_k; \psi)$ modality provides a very interesting new perspective on the logical analysis of neighborhood structures (beyond its use to find a recursion axioms for the evidence removal operation). See van Benthem et al. (2017, 2017a, 2017b) for further discussion and many interesting results using the $\Box(\varphi_1, \ldots, \varphi_k; \psi)$ modality. Among other results, van Benthem et al. develop an appropriate notion of bisimulation and provide a sound and complete axiomatization of neighborhood frames.

Returning to the question of a recursion axiom for the evidence removal modality, the new modality is exactly what is needed[18]:

Lemma 3.10 $[-\varphi]\langle\ ]\psi \leftrightarrow (\neg[A]\varphi \rightarrow \Box(\neg\varphi; [-\varphi]\psi)$ *is valid on all evidence models.*

Proof Let $\mathcal{M} = \langle W, E, V \rangle$ be an evidence model with $[\![\varphi]\!]_{\mathcal{M}} \neq W$ (otherwise, for all w, $E^{-\varphi}(w) = \emptyset$). We show that $[-\varphi]\langle\ ]\psi \leftrightarrow \Box_{\neg\varphi}[-\varphi]\psi$ is valid on $\mathcal{M}$. Let $w \in W$. The key observation is that

(∗)   for all $X \subseteq W$, $X \in E^{-\varphi}(w)$ iff $X \in E(w)$ and $X \cap [\![\neg\varphi]\!]_{\mathcal{M}} \neq \emptyset$.

Then, for each $w \in W$,

$\mathcal{M}, w \models [-\varphi]\langle\ ]\psi$ iff $\mathcal{M}^{-\varphi}, w \models \langle\ ]\psi$
     iff there is a $X \in E^{-\varphi}(w)$ such that $X \subseteq [\![\psi]\!]_{\mathcal{M}^{-\varphi}}$
     iff there is a $X \in E(w)$ such that $X \cap [\![\neg\varphi]\!]_{\mathcal{M}}$
     and $X \subseteq [\![[-\varphi]\psi]\!]_{\mathcal{M}}$
     (this follows from (∗) and $[\![\psi]\!]_{\mathcal{M}^{-\varphi}} = [\![[-\varphi]\psi]\!]_{\mathcal{M}}$)
     iff $\mathcal{M}, w \models \Box(\neg\varphi; [-\varphi]\psi)$.

Thus, $[-\varphi]\langle\ ]\psi \leftrightarrow (\neg[A]\varphi \rightarrow \Box(\neg\varphi; [-\varphi]\psi)$ is valid on all evidence models. □

Note how this principle captures the logical essence of evidence removal.

Recall the definition of the conditional evidence modality (Definition 3.31): An agent has evidence that ψ conditional on φ if there is evidence consistent with φ such that restriction to the worlds where φ is true entails ψ. Our next conditional operator $\Box_\varphi \psi$ drops the latter condition: it is true if the agent has evidence compatible

---

[18] The precondition is needed because the set of all worlds W is an evidence set.

with φ that entails ψ. In general, we include operators $\Box_{\overline{\varphi}}\psi$ where $\overline{\varphi}$ is a sequence of formulas. The intended interpretation is that "ψ is entailed by some admissible evidence *compatible* with each of $\overline{\varphi}$". After removing φ, there is evidence for ψ there is currently evidence for ψ that was not removed by the $-\varphi$ operation. The evidence that is not removed is the evidence that is compatible with $\neg\varphi$. This $\Box(\varphi; \psi)$ there is there evidence $X \in N(w)$ such that.

However, we are not done yet. We also need a recursion axiom for our new operator $\Box(\varphi_1, \ldots, \varphi_k; \psi)$. This, in turn, requires another extension to our language. The new modality combines the conditional neighborhood modality (Definition 3.31) with the modality from Definition 3.34.

Definition 3.35 (*Conditional instantial neighborhood modality*) Suppose that $\varphi_1, \ldots, \varphi_k, \alpha$ and ψ are formulas (e.g., from $\mathcal{L}$ or $\mathcal{L}^{ev}$). Then, $\Box^{\alpha}(\varphi_1, \ldots, \varphi_k; \psi)$ is a formula interpreted at states in a neighborhood (evidence) model $\mathcal{M} = \langle W, N, V \rangle$ as follows:

$\mathcal{M}, w \models \Box^{\alpha}(\varphi_1, \ldots, \varphi_k; \psi)$ iff there exists a set $X \in N(w)$ such that (i) for all $i = 1, \ldots, k$, (ii) $X \cap [\![\varphi_i]\!]_{\mathcal{M}} \neq \emptyset$, $X \cap [\![\alpha]\!]_{\mathcal{M}} \neq \emptyset$, and (iii) $X \cap [\![\alpha]\!]_{\mathcal{M}} \subseteq [\![\psi]\!]_{\mathcal{M}}$.

Note that this modality can define the instantial neighborhood modality defined above: $\Box(\varphi_1, \ldots, \varphi_k; \psi)$ is equivalent to $\Box^{\top}(\varphi_1, \ldots, \varphi_k; \psi)$. We are now ready to state the recursion axioms for evidence removal.

$$[-\varphi] p \leftrightarrow (\neg[A]\varphi \to p) \quad (p \in \mathsf{At})$$
$$[-\varphi](\psi \wedge \chi) \leftrightarrow ([-\varphi]\psi \wedge [-\varphi]\chi)$$
$$[-\varphi]\neg\psi \leftrightarrow (\neg[A]\varphi \to \neg[-\varphi]\psi)$$
$$[-\varphi]\Box^{\alpha}(\psi_1, \ldots, \psi_k; \chi) \leftrightarrow (\neg[A]\varphi \to \Box^{[-\varphi]\alpha}([-\varphi]\psi_1, \ldots, [-\varphi]\psi_k, \neg\varphi; [-\varphi]\chi))$$
$$[-\varphi][A]\psi \leftrightarrow (\neg[A]\varphi \to [A][-\varphi]\psi)$$

Consult van Benthem and Pacuit (2011) for further discussion (including an explanation of how to extend the analysis in the section to the full language involving the belief modalities).

The main conclusion from this section is that a logical analysis of evidence dynamics that changes neighborhood models uncovers the need for new modalities in our base language of evidence models. Clearly, there are many new questions of axiomatization resulting from this. For the modal logician, the pleasant surprise is that there is a lot of well-motivated new modal structure in neighborhood models that is waiting to be explored.

Appendix A
Relational Semantics for Modal Logic

This Appendix provides a very brief introduction to relational semantics for modal logic. The goal is to provide just enough details to motivate the discussion of neighborhood semantics and facilitate a comparison between the two semantics. There are many textbooks that you can consult for more information (for example, van Benthem 2010; Blackburn et al. 2001; Chellas 1980; Fitting and Mendelsohn 1999; Goldblatt 2011).[1]

Definition A.1 (*Relational Frame and Model*) A **relational frame** is a tuple $\langle W, R \rangle$ where W is a nonempty set (elements of W are called **states**) and $R \subseteq W \times W$ is a relation on W. A **relational model** (also called a Kripke model) is a triple $\mathfrak{M} = \langle W, R, V \rangle$ where $\langle W, R \rangle$ is a relational frame and $V : \mathsf{At} \to \wp(W)$ (At is the set of atomic propositions) is a **valuation function** assigning sets of states to atomic propositions.

Example A.2 The following picture represents the relational structure $\mathfrak{M} = \langle W, R, V \rangle$ where $W = \{w_1, w_2, w_3, w_4\}$, $R = \{(w_1, w_2), (w_1, w_3), (w_1, w_4), (w_2, w_2), (w_2, w_4), (w_3, w_4)\}$ and $V(p) = \{w_2, w_3\}$ and $V(q) = \{w_3, w_4\}$.

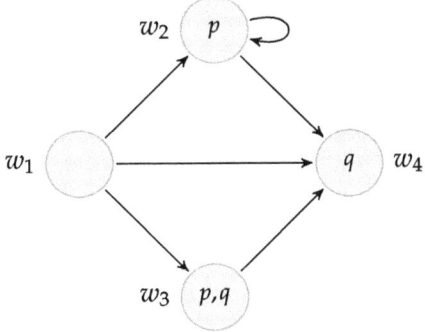

---

[1] This is not a complete list, but a pointer to books that covers topics related to issues discussed in this book. See Chagrov and Zakharyaschev (1997), Kracht (1999), Goldblatt (1992a), and Humberstone (2016) for different perspectives on modal logic.

Formulas of $\mathcal{L}(\mathsf{At})$ (Definition 1.6) are interpreted at states in a relational model.

Definition A.3 (*Truth of Modal Formulas*) Suppose that $\mathfrak{M} = \langle W, R, V \rangle$ is a relational model. Truth of a modal formula $\varphi \in \mathcal{L}(\mathsf{At})$ at a state w in $\mathfrak{M}$, denoted $\mathfrak{M}, w \models \varphi$, is defined inductively as follows:

1. $\mathfrak{M}, w \models p$ iff $w \in V(p)$ (where $p \in \mathsf{At}$).
2. $\mathfrak{M}, w \models \neg\varphi$ iff $\mathfrak{M}, w \not\models \varphi$.
3. $\mathfrak{M}, w \models \varphi \wedge \psi$ iff $\mathfrak{M}, w \models \varphi$ and $\mathfrak{M}, w \models \psi$.
4. $\mathfrak{M}, w \models \Box\varphi$ iff for all $v \in W$, if $w R v$ then $\mathfrak{M}, v \models \varphi$.

The definition of truth for the other Boolean connectives ($\rightarrow, \vee, \leftrightarrow$) is as usual. Let $\bot$ and $\top$ be atomic proposition with the fixed meaning: for all states w, $\mathfrak{M}, w \models \top$ and $\mathfrak{M}, w \not\models \bot$. Finally, assume that the '$\Diamond$' modality is defined as follows: $\Diamond\varphi$ is $\neg\Box\neg\varphi$. Then, the definition of truth for $\Diamond\varphi$ is:

- $\mathfrak{M}, w \models \Diamond\varphi$ iff there is a $v \in W$ such that $w R v$ and $\mathfrak{M}, v \models \varphi$.

Definition A.4 (*Truth Set*) Suppose that $\mathfrak{M} = \langle W, R, V \rangle$ is a relational model. For each $\varphi \in \mathcal{L}$, let $[\![\varphi]\!]_{\mathfrak{M}} = \{w \in W \mid \mathfrak{M}, w \models \varphi\}$ be the **truth set** of φ (in $\mathfrak{M}$).

For an extended discussion surrounding the interpretation modal formulas in relational models, see (van Benthem 2010, Chap. 2).

Exercise A.1 Consult http://pacuit.org/modal/tutorials/ for more examples to test your understanding of the definition of truth for modal formulas over relational models.

Definition A.5 (*Validity*) A modal formula $\varphi \in \mathcal{L}$ is **valid in a relational model** $\mathfrak{M} = \langle W, R, V \rangle$, denoted $\mathfrak{M} \models \varphi$, provided $\mathfrak{M}, w \models \varphi$ for each $w \in W$. Suppose that $\mathfrak{F} = \langle W, R \rangle$ is a relational frame. A modal formula $\varphi \in \mathcal{L}$ is valid on $\mathfrak{F}$, denoted $\mathfrak{F} \models \varphi$, provided $\mathfrak{M} \models \varphi$ for all models based on $\mathfrak{F}$ (i.e., all models $\mathfrak{M} = \langle \mathfrak{F}, V \rangle$). Suppose that F is a class of relational frames. A modal formula φ is **valid on** F, denoted $\models_\mathsf{F} \varphi$, provided $\mathfrak{F} \models \varphi$ for all $\mathfrak{F} \in \mathsf{F}$. If F is the class of all relational frames, then I will write $\models \varphi$ instead of $\models_\mathsf{F} \varphi$.

In order to show that a modal formula φ is valid, it is enough to argue informally that φ is true at an arbitrary state in an arbitrary relational model. On the other hand, to show a modal formula φ is not valid, one must provide a counter example (i.e., a relational model and state where an instance of φ is false).

A.1 Definability

Definition A.4 explains how to assign to every modal formula $\varphi \in \mathcal{L}$ a set of states in a relational model $\mathfrak{M} = \langle W, R, V \rangle$. It is natural to ask about the converse: Given and arbitrary set, when does a formula uniquely pick out that set?

Appendix A: Relational Semantics for Modal Logic

Definition A.6 (*Definable Subsets*) Let $\mathfrak{M} = \langle W, R, V \rangle$ be a relational model. A set $X \subseteq W$ is **definable in** $\mathfrak{M}$ provided $X = [\![\varphi]\!]_{\mathfrak{M}}$ for some modal formula $\varphi \in \mathcal{L}$.

Example A.7 All four of the states in the relational model below are defined by a modal formula:

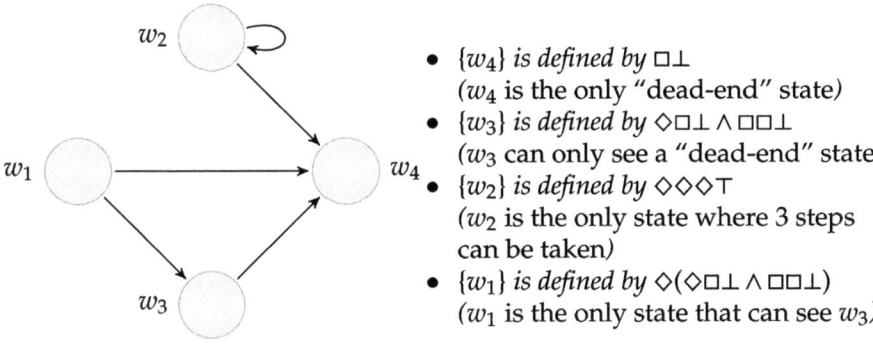

- $\{w_4\}$ *is defined by* $\Box\bot$
  (w_4 is the only "dead-end" state)
- $\{w_3\}$ *is defined by* $\Diamond\Box\bot \wedge \Box\Box\bot$
  (w_3 can only see a "dead-end" state)
- $\{w_2\}$ *is defined by* $\Diamond\Diamond\Diamond\top$
  (w_2 is the only state where 3 steps can be taken)
- $\{w_1\}$ *is defined by* $\Diamond(\Diamond\Box\bot \wedge \Box\Box\bot)$
  (w_1 is the only state that can see w_3)

Note that even in finite relational models, not all subsets may be definable. A problem can arise if states cannot be distinguished by modal formulas. For example, if the reflexive arrow is dropped in the relational model above, then w_2 and w_3 *cannot* be distinguished by a modal formula (there are ways to formally prove this, but see if you can informally argue why w_2 and w_3 cannot be distinguished in this case).

The next two definitions make precise what it means for two states to be *indistinguishable* by a modal formula.

Definition A.8 (*Modal Equivalence*) Suppose that $\mathfrak{M}_1 = \langle W_1, R_1, V_1 \rangle$ and $\mathfrak{M}_2 = \langle W_2, R_2, V_2 \rangle$ are two relational models. We say $\mathfrak{M}_1, w_2$ and $\mathfrak{M}_2, w_2$ are **modally equivalent** provided that

$$\text{for all modal formulas } \varphi \in \mathcal{L}, \ \mathfrak{M}_1, w_1 \models \varphi \text{ iff } \mathfrak{M}_2, w_2 \models \varphi.$$

We write $\mathfrak{M}_1, w_1 \leftrightsquigarrow \mathfrak{M}_2, w_2$ if $\mathfrak{M}_1, w_1$ and $\mathfrak{M}_2, w_2$ are modally equivalent. (Note that it is assumed $w_1 \in W_1$ and $w_2 \in W_2$)

Definition A.9 (*Bisimulation*) Let $\mathfrak{M}_1 = \langle W_1, R_1, V_1 \rangle$ and $\mathfrak{M}_2 = \langle W_2, R_2, V_2 \rangle$ be two relational models. A nonempty relation $Z \subseteq W_1 \times W_2$ is called a **bisimulation** provided for all $w_1 \in W_1$ and $w_2 \in W_2$, if $w_1 Z w_2$ then

1. (atomic harmony) For all $p \in \mathsf{At}$, $w_1 \in V_1(p)$ iff $w_2 \in V_2(p)$.
2. (zig) If $w_1 R_1 v_1$ then there is a $v_2 \in W_2$ such that $w_2 R_2 v_2$ and $v_1 Z v_2$.
3. (zag) If $w_2 R_2 v_2$ then there is a $v_1 \in W_1$ such that $w_1 R_1 v_1$ and $v_1 Z v_2$.

We write $\mathfrak{M}_1, w_1 \underline{\leftrightarrow} \mathfrak{M}_2, w_2$ if there is a bisimulation relating w_1 with w_2.

Definitions A.8 and A.9 provide two concrete ways to answer the question: *when are two states the same?*

Example A.10 (*Bisimulation Example*) The dashed lines is a bisimulation between the following two relational models (for simplicity, we assume that all atomic propositions are false):

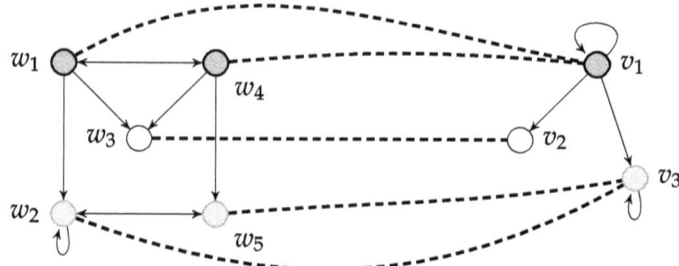

On the other hand, there is no bisimulation relating the states x and y in the following two relational models:

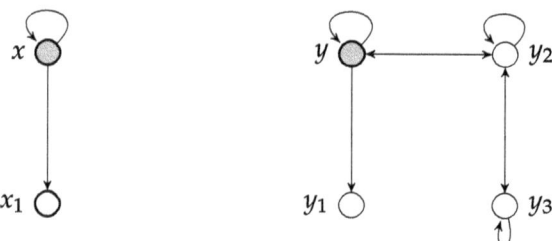

Using Lemma A.1 below, we can *prove* that there is no bisimulation relating x and y. We first note that $\Box(\Diamond\Box\bot \vee \Box\bot)$ is true at state x but not true at state y. Then by Lemma A.1, x and y cannot be bisimilar.

Lemma A.1 (Modal Invariance Lemma) *Suppose* $\mathfrak{M}_1 = \langle W_1, R_1, V_1 \rangle$ *and* $\mathfrak{M}_2 = \langle W_2, R_2, V_2 \rangle$ *are relational models. For all* $w \in W_1$ *and* $v \in W_2$, *if* $\mathfrak{M}_1, w \underline{\leftrightarrow} \mathfrak{M}_2, v$ *then* $\mathfrak{M}_1, w \rightsquigarrow \mathfrak{M}_2, v$.

Lemma A.2 *Suppose* $\mathfrak{M}_1 = \langle W_1, R_1, V_1 \rangle$ *and* $\mathfrak{M}_2 = \langle W_2, R_2, V_2 \rangle$ *are finite relational models. If* $\mathfrak{M}_1, w_1 \rightsquigarrow \mathfrak{M}_2, w_2$ *then* $\mathfrak{M}_1, w_1 \underline{\leftrightarrow} \mathfrak{M}_2, w_2$.

Proof We show that $\rightsquigarrow$ is a bisimulation. The atomic harmony condition is obvious. We prove the zag condition. Suppose that $\mathfrak{M}_1, w_1 \rightsquigarrow \mathfrak{M}_2, w_2$, $w_2 R_2 v_2$, but there is no v_1 such that $w_1 R_1 v_1$ and $\mathfrak{M}_1, v_1 \rightsquigarrow \mathfrak{M}_2, v_2$. Note that there are only finitely many states that are accessible from w_1. That is, $\{w \mid w_1 R_1 w\}$ is a finite set. Suppose that $\{w \mid w_1 R_1 w\} = \{w^1, w^2, \ldots, w^m\}$. By assumption, for each w^i we have $\mathfrak{M}_1, w^i \not\rightsquigarrow \mathfrak{M}_2, v_2$. Hence, for each w^i, there is a formula φ_i such that $\mathfrak{M}_1, w^i \not\models \varphi_i$ but $\mathfrak{M}_2, v_2 \models \varphi_i$. Then, $\mathfrak{M}_2, v_2 \models \bigwedge_{i=1,\ldots,m} \varphi_i$. Since $w_2 R_2 v_2$, we have $\mathfrak{M}_2, w_2 \models \Diamond \bigwedge_{i=1,\ldots,m} \varphi_i$. Therefore, $\mathfrak{M}_1, w_1 \models \Diamond \bigwedge_{i=1,\ldots,m} \varphi_i$. But this is a contradiction, since the only states accessible from w_1 are $w^1, \ldots, w^m$, and for each w^i there is a φ_i such that $\mathfrak{M}_1, w^i \not\models \varphi_i$. The proof of the zag condition is similar. $\square$

The modal invariance Lemma (Lemma A.1) can be used to prove what can and cannot be expressed in the basic modal language.

Appendix A: Relational Semantics for Modal Logic   143

Fact A.11 *Let $\mathfrak{M} = \langle W, R, V \rangle$ be a relational model. The "exists two" operator $\Diamond_2 \varphi$ is: $\mathfrak{M}, w \models \Diamond_2 \varphi$ iff there is $v_1, v_2 \in W$ such that $v_1 \neq v_2$, $\mathfrak{M}, v_1 \models \varphi$ and $\mathfrak{M}, v_2 \models \varphi$. The exist two operator $\Diamond_2$ is not definable in the basic modal language.*

Proof Suppose that the $\Diamond_2$ is definable in the basic modal language. Then there is a basic modal formula $\alpha(\cdot)$ such that for any formula φ and any relational model $\mathfrak{M}$ with state w, we have $\mathfrak{M}, w \models \Diamond_2 \varphi$ iff $\mathfrak{M}, w \models \alpha(\varphi)$. Consider the relational model $\mathfrak{M} = \langle W, R, V \rangle$ with $W = \{w_1, w_2, w_3\}$, $R = \{(w_1, w_2), (w_1, w_3)\}$ and $V(p) = \{w_2, w_3\}$. Note that $\mathfrak{M}, w_1 \models \Diamond_2 p$. Since $\Diamond_2$ is assumed to be defined by $\alpha(\cdot)$, we must have $\mathfrak{M}, w_1 \models \alpha(p)$. Consider the relational model $\mathfrak{M}' = \langle W', R', V' \rangle$ with $W' = \{v_1, v_2\}$, $R' = \{(v_1, v_2)\}$ and $V'(p) = \{v_2\}$. Note that $Z = \{(w_1, v_1), (w_2, v_2), (w_3, v_2)\}$ is a bisimulation relating w_1 and v_1 (i.e., $\mathfrak{M}, w_1 \underline{\leftrightarrow} \mathfrak{M}', v_1$). By Lemma A.1, $\mathfrak{M}, w_1 \leftrightsquigarrow \mathfrak{M}', v_1$. Therefore, since $\alpha(p)$ is a formula of the basic modal language and $\mathfrak{M}, w_1 \models \alpha(p)$, we have $\mathfrak{M}', v_1 \models \alpha(p)$. Since $\alpha(\cdot)$ defines $\Diamond_2$, $\mathfrak{M}', v_1 \models \Diamond_2 p$, which is a contradiction. Hence, $\Diamond_2$ is not definable in the basic modal language. □

The basic modal language can also be used to define *classes* of relational frames. Suppose that P is a property of relations (eg., reflexivity or transitivity). We say a frame $\mathfrak{F} = \langle W, R \rangle$ has property P provided R has property P. For example, $\mathfrak{F} = \langle W, R \rangle$ is called a **transitive frame** provided R is transitive, i.e., for all $w, x, v \in W$, if wRx and xRv then wRv.

Definition A.12 (*Defining a Class of Frames*) A modal formula φ **defines the class of frames with property** P provided for all frames $\mathfrak{F}$, $\mathfrak{F} \models \varphi$ iff $\mathfrak{F}$ has property P.

Remark A.13 (*Remark on validity on frames*) Note that if $\mathfrak{F} \models \varphi$ where φ is some modal formula, then $\mathfrak{F} \models \varphi^*$ where φ^* is any **substitution instance** of φ. That is, φ^* is obtained by replacing sentence letters in φ with modal formulas. In particular, this means, for example, that in order to show that $\mathfrak{F} \not\models \Box \varphi \to \varphi$ it is enough to show that $\mathfrak{F} \not\models \Box p \to p$ where p is a sentence letter.

Fact A.14 $\Box \varphi \to \Box \Box \varphi$ *defines the class of transitive frames.*

Proof We must show for any frame $\mathfrak{F}$, $\mathfrak{F} \models \Box \varphi \to \Box \Box \varphi$ iff $\mathfrak{F}$ is transitive.
($\Leftarrow$) Suppose that $\mathfrak{F} = \langle W, R \rangle$ is transitive and let $\mathfrak{M} = \langle W, R, V \rangle$ be any model based on $\mathfrak{F}$. Given $w \in W$, we must show $\mathfrak{M}, w \models \Box \varphi \to \Box \Box \varphi$. Suppose that $\mathfrak{M}, w \models \Box \varphi$. We must show $\mathfrak{M}, w \models \Box \Box \varphi$. Suppose that $v \in W$ and wRv. We must show $\mathfrak{M}, v \models \Box \varphi$. To that end, let $x \in W$ be any state with vRx. Since R is transitive and wRv and vRx, we have wRx. Since $\mathfrak{M}, w \models \Box \varphi$, we have $\mathfrak{M}, x \models \varphi$. Therefore, since x is an arbitrary state accessible from v, $\mathfrak{M}, v \models \Box \varphi$. Hence, $\mathfrak{M}, w \models \Box \Box \varphi$, and so, $\mathfrak{M}, w \models \Box \varphi \to \Box \Box \varphi$, as desired.

($\Rightarrow$, *by contraposition*) We argue by contraposition. Suppose that $\mathfrak{F}$ is not transitive. We must show $\mathfrak{F} \not\models \Box \varphi \to \Box \Box \varphi$. By the above Remark, it is enough to show $\mathfrak{F} \not\models \Box p \to \Box \Box p$ for some sentence letter p. Since $\mathfrak{F}$ is not transitive, there are states $w, v, x \in W$ with wRv and vRx but it is not the case that wRx. Consider the model $\mathfrak{M} = \langle W, R, V \rangle$ based on $\mathfrak{F}$ with $V(p) = \{y \mid y \neq x\}$. Since $\mathfrak{M}, x \not\models p$ and

wRv and vRx, we have $\mathfrak{M}, w \not\models \Box\Box p$. Furthermore, $\mathfrak{M}, w \models \Box p$ since the only state where p is false is x and it is assumed that it is not the case that wRx. Therefore, $\mathfrak{M}, w \models \Box p \land \neg\Box\Box p$, and so, $\mathfrak{F} \not\models \Box p \to \Box\Box p$, as desired. □

A.2 Normal Modal Logics

The definition of the minimal normal modal logic **K** is given in Sect. 2.3. Recall the definition of a deduction $\vdash_L \varphi$ (Definition 2.25) and deduction from assumptions $\Gamma \vdash_L \varphi$ (Definition 2.30). The following is a deduction showing that $\vdash_K \Box(\varphi \land \psi) \to (\Box\varphi \land \Box\psi)$:

1. $(\varphi \land \psi) \to \varphi$ instance of a tautology
2. $\Box((\varphi \land \psi) \to \varphi)$ (Nec) 1
3. $\Box((\varphi \land \psi) \to \varphi) \to (\Box(\varphi \land \psi) \to \Box\varphi)$ instance of (K)
4. $\Box(\varphi \land \psi) \to \Box\varphi$ (MP) 2,3
5. $(\varphi \land \psi) \to \psi$ instance of a tautology
6. $\Box((\varphi \land \psi) \to \psi)$ (Nec) 5
7. $\Box((\varphi \land \psi) \to \varphi) \to (\Box(\varphi \land \psi) \to \Box\psi)$ instance of (K)
8. $\Box(\varphi \land \psi) \to \Box\psi$ (MP) 5,6
9. $(a \to b) \to ((a \to c) \to (a \to (b \land c)))$ tautology ($a := \Box(\varphi \land \psi)$, $b := \Box\varphi, c := \Box\psi$)
10. $(a \to c) \to (a \to (b \land c))$ (MP) 4,9
11. $\Box(\varphi \land \psi) \to \Box\varphi \land \Box\psi$ (MP) 8,10

The following axiom schemes have played an important role in both the mathematical development and applications of modal logic.

(K) $\Box(\varphi \to \psi) \to (\Box\varphi \to \Box\psi)$ (4) $\Box\varphi \to \Box\Box\varphi$
(D) $\Box\varphi \to \Diamond\varphi$ (5) $\neg\Box\varphi \to \Box\neg\Box\varphi$
(T) $\Box\varphi \to \varphi$ (L) $\Box(\Box\varphi \to \varphi) \to \Box\varphi$

Each of the above formulas are called **axiom schemas** and I will often refer to **instances** of these axiom schemas. The general idea is to treat the 'φ' in the above formulas as a meta-variable that can be replaced by specific formulas from $\mathcal{L}$. For instance, $\Box\Diamond p \to \Diamond p$ is a substitution instance of the axiom scheme (T).

Recall from Sect. 2.3 that the minimal normal modal logic, **K**, is the smallest set of formulas that contains all tautologies, all instances of K, all instances of (Dual), and is closed under the rules (Nec) (from φ infer $\Box\varphi$) and (MP) (from φ and $\varphi \to \psi$ infer ψ). Other normal modal logics are defined by adding all instances of axiom schema or rules to **K**. If $A_1, \ldots, A_n$ are axiom schemas, then write $\mathbf{K} + A_1 + A_2 + \cdots + A_n$

for the extension of **K** with the axioms $A_1, \ldots, A_n$. Using this naming convention for logics, I can now define a number of well-studied normal modal logics:

$$\begin{aligned} \mathbf{T} &\text{ is } \mathbf{K} + (\mathsf{T}) \\ \mathbf{S4} &\text{ is } \mathbf{K} + (\mathsf{T}) + (4) \\ \mathbf{S5} &\text{ is } \mathbf{K} + (\mathsf{T}) + (4) + (5) \\ \mathbf{KD45} &\text{ is } \mathbf{K} + (\mathsf{D}) + (4) + (5) \\ \mathbf{GL} &\text{ is } \mathbf{K} + (\mathsf{L}) \end{aligned}$$

Definition A.15 (*Semantic Consequence*) Suppose that Γ is a set of modal formulas and F is a class of relational frames. We say φ is a **semantic consequence** of Γ with respect to F, denoted $\Gamma \models_\mathsf{F} \varphi$, provided for all models $\mathfrak{M} = \langle W, R, V \rangle$ based on a frame from F (i.e., $\langle W, R \rangle \in \mathsf{F}$) and all states $w \in W$, if $\mathfrak{M}, w \models \Gamma$, then $\mathfrak{M}, w \models \varphi$ (where $\mathfrak{M}, w \models \Gamma$ when $\mathfrak{M}, w \models \gamma$ for all $\gamma \in \Gamma$).

Definition A.16 (*Soundness, Weak/Strong Completeness*) Suppose that F is a class of relational frames. A logic **L** is **sound** with respect to F provided, for all sets of formulas Γ, if $\Gamma \vdash_\mathbf{L} \varphi$, then $\Gamma \models_\mathsf{F} \varphi$. A logic **L** is **strongly complete** with respect to F provided for all sets of formulas Γ, if $\Gamma \models_\mathsf{F} \varphi$, then $\Gamma \vdash_\mathbf{L} \varphi$. A logic **L** is **weakly complete** with respect to F provided if $\models_\mathsf{F} \varphi$, then $\vdash_\mathbf{L} \varphi$.

Clearly, if a logic is strongly complete then it is weakly complete. Interestingly, the converse is not true (as we will see below). The proofs of the following theorem can be found in Blackburn et al. (2001). Details of the technique used to prove strong completeness is discussed in Sect. 2.3.2.

Theorem A.17 (Completeness Theorems)

- **K** *is sound and strongly complete with respect to the class of all relational frames.*
- **T** *is sound and strongly complete with respect to the class of reflexive relational frames.*
- **S4** *is sound and strongly complete with respect to the class of reflexive and transitive relational frames.*
- **S5** *is sound and strongly complete with respect to the class of reflexive, transitive and Euclidean relational frames (i.e., relations that form a partition).*
- **KD45** *is sound and strongly complete with respect to the class of serial, transitive and Euclidean relational frames (i.e., relations that form a* quasi-partition).

The logic **GL** does not follow the same pattern as the logics mentioned in the above theorem. There is a natural class of relational frames that characterizes **GL**. A relation $R \subseteq W \times W$ is **converse well-founded** (also called Noetherian) if there is no infinite ascending chain of states—i.e., there is no infinite set of distinct elements $w_0, w_1, \ldots$ from W, such that $w_0 \, R \, w_1 \, R \, w_2 \cdots$. Note that if R is converse well-founded, then it is **irreflexive** (for all $w \in W$, $w \, \not\!R \, w$). It is not hard to see that **G** is sound with respect to the class of frames that are transitive and converse well-founded. However, **GL** is *not* strongly complete with respect to this class of frames. To see this, we need some additional notation.

Definition A.18 (*Compactness*) Suppose that **L** is sound with respect to some class of frames **F**. We say that **L** is **compact** provided that for any set of formulas Γ, if Γ is finitely satisfiable (every finite subset of formulas is satisfiable), then Γ is satisfiable.

Proposition A.1 *If* **L** *is sound and strongly complete with respect to some class of frames* **F**, *then* **L** *is compact.*

Proof Suppose that **L** is sound and strongly complete with respect to some class of frames **F**. Suppose that Γ is any set of formulas that is finitely satisfiable. I.e., every finite subset $\Gamma_0 \subseteq \Gamma$ has a model (based on a frame from **F**). If Γ is not satisfiable, then, since every consistent set is satisfiable, Γ is inconsistent. I.e., $\Gamma \vdash_\mathbf{L} \bot$. This means that there is a deduction from Γ in **L** of $\bot$. Since deductions are finite in length, only finitely many assumptions from Γ can be used in the deduction. This means that there is a finite subset $\Gamma_0 \subseteq \Gamma$ such that $\Gamma_0 \vdash_\mathbf{L} \bot$. By soundness, this means that Γ_0 is not satisfiable. This contradicts our assumption. Thus Γ is satisfiable. □

Observation A.19 *The logic* **GL** *is not strongly complete with respect to the class of transitive and converse well-founded relational frames.*

Nonetheless, Segerberg (1971) proved a *weak* completeness theorem for **GL**. The proof is beyond the scope of this Appendix (see Blackburn et al. 2001 for the details).

Theorem A.20 *The logic* **GL** *is sound and weakly complete with respect to the class of transitive and converse well-founded frames.*

References

Abdou, J. and H. Keiding (1991). *Effectivity Functions in Social Choice*. Kluwer Academic Publishers.

Alchourrón, C. E., P. Gärdenfors, and D. Makinson (1985). On the logic of theory change: Partial meet contraction and revision functions. *Journal of Symbolic Logic 50*, 510 – 530.

Allen, M. (2005). Complexity results for logics of local reasoning and inconsistent belief. In *Proceedings of the 10th TARK Conference*, pp. 92–108.

Andreka, H., M. Ryan, and P. Y. Schobbens (2002). Operators and laws for combining preference relations. *Journal of Logic and Computation 12*(1), 13–53.

Areces, C. and D. Figueira (2009). Which semantics for neighbourhood semantics? In *Proceedings of the International Joint Conference on Artificial Intelligence*, Pasadena, California, USA, pp. 671 – 676.

Areces, C. and B. ten Cate (2007). Hybrid logics. In *Handbook of Modal Logic*, Volume 3, pp. 821 – 868. Elsevier.

Arló-Costa, H. (2002). First order extensions of classical systems of modal logic. *Studia Logica 71*, 87 – 118.

Arló-Costa, H. (2005). Non-adjunctive inference and classical modalities. *Journal of Philosophical Logic 34*(5-6), 581–605.

Arló-Costa, H. (2007). The logic of conditionals. In E. N. Zalta (Ed.), *The Stanford Encyclopedia of Philosophy* (Winter 2016 ed.).

Arló-Costa, H. and E. Pacuit (2006). First-order classical modal logic. *Studia Logica 84*, 171 – 210.

Artemov, S. and T. Protopopescu (2016). Intuitionistic epistemic logic. *Review of Symbolic Logic 9*, 266 – 298.

Awodey, S. and K. Kishida (2008). Topology and modality: the topological interpretation of first-order modal logic. *Review of Symbolic Logic 1*, 146 – 166.

Baltag, A., N. Bezhanishvili, A. Özgün, and S. Smets (2013). The topology of belief, belief reivsion and defeasible knowledge. In D. Grossi, O. Roy, and H. Huang (Eds.), *Logic, Rationality, and Interaction, 4th Workshop LORI 2013, Hangzhou, China*, pp. 27 – 40.

Baltag, A., N. Bezhanishvili, A. Özgün, and S. Smets (2015). The topological theory of belief. Technical report, ILLC preprint PP-2015-18.

Baltag, A. and S. Smets (2006a). Conditional doxastic models: A qualitative approach to dynamic belief revision. In G. Mints and R. de Queiroz (Eds.), *Proceedings of WOLLIC 2006, LNCS*, Volume 165, pp. 5–21.

Baltag, A. and S. Smets (2006b). Dynamic belief revision over multi-agent plausibility models. In G. Bonanno, W. van der Hoek, and M. Wooldridge (Eds.), *Proceedings of the 7th Conference on Logic and the Foundations of Game and Decision (LOFT 2006)*, pp. 11–24.

Barwise, J. (1987). Three views of common knowledge. In *Proceedings of Theoretical Aspects of Rationality and Knowledge (TARK)*.

Beklemishev, L. and D. Gabelaia (2014). Topological interpretations of provability logic. In *Leo Esakia on Duality in Modal and Intuitionistic Logic*, pp. 257 – 290.

van Benthem, J. (1978). Two simple incomplete modal logics. *Theoria 44*, 25 – 37.

van Benthem, J. (2003). Logic games are complete for game logics. *Studia Logica 75*(2), 183 – 203.

van Benthem, J. (2004). Dynamic logic for belief revision. *Journal of Applied Non-Classical Logics 14*(2), 129–155.

van Benthem, J. (2010). *Modal Logic for Open Minds*. CSLI Publications.

van Benthem, J. (2011). *Logical Dynamics of Information and Interaction*. Cambridge University Press.

van Benthem, J. (2014). *Logic in Games*. The MIT Press.

van Benthem, J., N. Bezhanishvili, and S. Enqvist (2017a). A new game equivalence and its modal logic. In *Proceedings of Theoretical Aspects of Rationality and Knowledge (TARK 2017)*.

van Benthem, J., N. Bezhanishvili, and S. Enqvist (2017b). A propositional dynamic logic for instantial neighborhood models. johan van benthem, nick bezhanishvili, sebastian enqvist. proceedings of. In *Proceedings of Logics and Rational Interaction (LORI VI)*.

van Benthem, J., N. Bezhanishvili, S. Enqvist, and J. Yu (2017). Instantial neighbourhood logic. *Review of Symbolic Logic 10*(1), 116 – 144.

van Benthem, J. and G. Bezhanisvilli (2007). Modal logics of space. In *Handbook of Spatial Logic*, pp. 217 – 298. Springer.

van Benthem, J., D. Fernández-Duque, and E. Pacuit (2012). Evidence logic: A new look at neighborhood structures. In *Advances in Modal Logic*. College Publications.

van Benthem, J., D. Fernández-Duque, and E. Pacuit (2014). Evidence and plausibility in neighborhood structures. *Annals of Pure and Applied Logic 165*(1), 106–133.

van Benthem, J., S. Ghosh, and F. Liu (2008). Modelling simultaneous games with dynamic logic. *Synthese 165*(2), 247 – 268.

van Benthem, J., D. Grossi, and F. Liu (2014). Priority structures in deontic logic. *Theoria 80*(2), 116 – 152.

van Benthem, J. and E. Pacuit (2011). Dynamic logics of evidence-based beliefs. *Studia Logica 99*(1-3), 61–92.

van Benthem, J. and D. Sarenac (2004). The geometry of knowledge. In *Aspects of Universal Logic*, pp. 1 – 31.

Benton, R. (2002). A simple incomplete extension of **T** which is the union of two complete modal logics with f.m.p. *Journal of Philosophical Logic 31*(6), 527 – 541.

Berto, F. (2013). Impossible worlds. In E. N. Zalta (Ed.), *The Stanford Encyclopedia of Philosophy* (Winter 2013 ed.).

Berwanger, D. (2003). Game logic is strong enough for parity games. *Studia Logica 75*(2), 205 – 219.

Bezhanishvili, G., L. Esakia, and D. Gabelaia (2010). The modal logic of stone space: Diamond as derivative. *The Review of Symbolic Logic 3*(1), 26 – 40.

Bezhanishvili, G., D. Gabelaia, and J. Lucero-Bryan (2015). Modal logics of metric spaces. *Review of Symbolic Logic 8*(1), 178 – 192.

Bezhanishvili, G. and M. Gehrke (2005). Completeness of S4 with respect to the real line: Revisited. *Annals of Pure and Applied Logic 131*, 287 – 301.

Bjorndahl, A. (2016). Topological subset space models for public announcements. In *Jaakko Hintikka Volume in Outstanding Contributions to Logic*.

Blackburn, P., M. de Rijke, and Y. Venema (2001). *Modal Logic*, Volume 58 of *Cambridge Tracts in Theoretical Computer Science*. Cambridge University Press.

References

Board, O. (2004). Dynamic interactive epistemology. *Games and Economic Behavior 49*, 49 – 80.
Bonanno, G. (1992). Set-theoretic equivalence of extensive-form games. *International Journal of Game Theory 20*(4), 429 – 447.
Boolos, G. and G. Sambin (1985). An incomplete system of modal logic. *Journal of Philosophical Logic 14*(4), 351 – 358.
Boutilier, C. (1992). *Conditional Logics for Default Reasoning and Belief Revision*. Ph. D. thesis, University of Toronto.
Bretto, A. (2013). *Hypergraph Theory: An Introduction*. Springer.
Calardo, E. (2013). *Non-normal Modal Logics, and Deontic Dilemmas: A Study in Multi-relational Semantics*. Ph. D. thesis, Universita Di Bologna.
Carr, D. (1979). The logic of knowing how and ability. *Mind 88*, 394 – 409.
Chagrov, A. and M. Zakharyaschev (1997). *Modal Logic*. Oxford University Press.
Chellas, B. (1980). *Modal Logic: An Introduction*. Cambridge: Cambridge University Press.
Chwe, M. (2001). *Rational Ritual: Culture, Coordination and Common Knowledge*. Princeton University Press.
Conradie, W., V. Goranko, and D. Vakarelov (2006). Algorithmic correspondence and completeness in modal logic I. the core algorithm SQEMA. *Logical Methods in Computer Science 2*(1-5), 1–26.
Cubitt, R. and R. Sugden (2003). Common knowledge, salience and convention: A reconstruction. *Economics and Philosophy 19*(2), 175 – 210.
Daniëls, T. (2011). Social choice and the logic of simple games. *Journal of Logic and Computation 21*(6), 883 – 906.
Dietrich, F. and C. List (2013). A reason-based theory of rational choice. *Nous 47*(1), 104 – 134.
van Ditmarsch, H., S. Knight, and A. Özgün (2015). Announcement as effort on topological spaces. In *Proceedings of TARK 2015*, pp. 95 – 102.
Došen, K. (1989). Duality between modal algebras and neighbourhood frames. *Studia Logica 48*(2), 219–234.
Dugundji, J. (1966). *Topology*. Wm. C. Brown Publishers.
Enderton, H. (2001). *A Mathematical Introduction to Logic*. Academic Press.
Fagin, R., J. Halpern, Y. Moses, and M. Vardi (1995). *Reasoning about Knowledge*. The MIT Press.
Fine, K. (1974). An incomplete logic containing S4. *Theoria 40*, 23–29.
Fitch, F. B. (1948). Intuitionistic modal logic with quantifiers. *Portugaliae Mathematicae 7*, 113 – 118.
Fitting, M. (1983). *Proof Methods for Modal and Intuitionistic Logics*. Reidel Publishing Co.
Fitting, M. (2006). Modal proof theory. In P. Blackburn, J. van Benthem, and F. Wolter (Eds.), *Handbook of Modal Logic*.
Fitting, M. and R. L. Mendelsohn (1999). *First-Order Modal Logic*. Synthese Library. Springer.
Forrester, J. W. (1984). Gentle murder, or the adverbial samaritan. *Journal of Philosophy 81*(4), 193 – 196.
Gabbay, D., V. Shehtman, and D. Skvortsov (2008). *Quantification in Nonclassical Logic*. Elsevier.
Gärdenfors, P. (1988). *Knowledge in Flux: Modeling the Dynamics of Epistemic States*. Bradford Books, MIT Press.
Garson, J. (2002). Quantification in modal logic. In D. Gabbay and F. Guenthner (Eds.), *Handbook of Philosophical Logic*, pp. 249 – 307.
Gasquet, O. and A. Herzig (1996). From classical to normal modal logic. In H. Wansing (Ed.), *Proof Theory of Modal Logic*. Kluwer.
Geanakoplos, J. (1995). Common knowledge. In R. J. Aumann and S. Hart (Eds.), *Handbook of Game Theory with Economic Applications*. Elsevier North-Holland.
Georgatos, K. (1993). *Modal Logics for Topological Spaces*. Ph. D. thesis, CUNY Graduate Center.
Georgatos, K. (1994). Knowledge theoretic properties of topological spaces. In M. Masuch and L. Polos (Eds.), *Lecture Notes in Artificial Intelligence*, Volume 808, pp. 147–159. Springer-Verlag.
Gerbrandy, J. (1999). *Bisimulations on Planet Kripke*. Ph. D. thesis, Institute for Logic, Language and Computation (DS-1999-01).

Gerson, M. (1975a). An extension of S4 complete for neighborhood semantics but incomplete for the relational semantics. *Studia Logica 34*, 333 — 342.

Gerson, M. (1975b). The inadequacy of the neighbourhood semantics for modal logics. *Journal of Symbolic Logic 40*, 141 – 148.

Gerson, M. (1976). A neighborhood frame for T with no equivalent relational frame. *Zeitschrift fur mathematische logik und Grundlagen der Mathematik 22*.

Gilbert, D. and P. Maffezioli (2015). Modular sequent calculi for classical modal logics. *Studia Logica 103*(1), 175 – 217.

Girard, P. (2008). *Modal Logic for Belief and Preference Change*. Ph. D. thesis, Stanford University, available at the ILLC University of Amsterdam Dissertation Series DS-2008-04.

Girlando, M., S. N. N. Olivetti, and V. Risch (2016). The logic of conditional beliefs: neighbourhood semantics and sequent calculus. In *Advances in Modal Logic*.

Goble, L. (1991). Murder most gentle: The paradox deepens. *Philosophical Studies 64*(2), 217–227.

Goble, L. (2000). Multiplex semantics for deontic logic. *Nordic Journal of Philosophical Logic 5*(2), 113 – 134.

Goble, L. (2004). A proposal for dealing with deontic dilemmas. In *Proceedings of the 7th Int. Workshop on Deontic Logic in Computer Science (DEON 2004)*, Volume 3065 of *Lecture Notes in Computer Science*. Springer.

Goldblatt, R. (1992a). *Logics of Time and Computation*. CSLI Publications.

Goldblatt, R. (1992b). Parallel action: Concurrent dynamic logic with independent modalities. *Studia Logica 51*, 551 – 578.

Goldblatt, R. (2011). *Quantifiers, Propositions and Identity: Admissible Semantics for Quantified Modal and Substructural Logics*. Lecture Notes in Logic. Cambridge University Press.

Goranko, V., W. Jamroga, and P. Turrini (2013). Strategic games and truly playable effectivity functions. *Autonomous Agents and Multi-Agent Systems 26*(2), 288 – 314.

Goranko, V. and S. Passy (1992). Using the universal modality: Gains and questions. *Journal of Logic and Computation 2*(1), 5 – 30.

Governatori, G. and A. Luppi (2000). Labelled tableaux for non-normal modal logics. In E. Lamma and P. Mello (Eds.), *AI*IA 99: Advances in Artificial Intelligence.*, pp. 119 – 130.

Governatori, G. and A. Rotolo (2005). On the axiomatization of elgesem's logic of agency and ability. *Journal of Philosophical Logic 34*(4), 403–431.

Grossi, D. and G. Pigozzi (2014). *Judgment Aggregation: A Primer*. Morgan & Claypool Publishers.

Grove, A. (1988). Two modellings for theory change. *Journal of Philosophical Logic 17*(2), 157 – 170.

Hakli, R. and S. Negri (2011). Does the deduction theorem fail for modal logic? *Synthese 187*(3), 849 – 867.

Halpern, J. and Y. Moses (1990). Knowledge and common knowledge in a distributed environment. *Journal of the ACM 37*(3), 549 – 587.

Halpern, J. and R. Pucella (2011). Dealing with logical omniscience: Expressiveness and pragmatics. *Artificial Intelligence 175*(1), 220 – 235.

Halpern, J. and L. Rêgo (2007). Characterizing the NP-PSPACE gap in the satisfiability problem for modal logic. *Journal of Logic and Computation 17*(4), 795 – 806.

Hansen, H. and C. Kupke (2004). A coalgebraic perspective on monotone modal logic. In *Proceedings of the 7th Workshop on Coalgebraic Methods in Computer Science (CMCS)*, Volume 106 of *Electronic Notes in Computer Science*, pp. 121–143. Elsevier.

Hansen, H. H. (2003). Monotonic modal logic (Master's thesis). Research Report PP-2003-24, ILLC, University of Amsterdam.

Hansen, H. H., C. Kupke, and E. Pacuit (2009). Neighbourhood structures: Bisimilarity and basic model theory. *Logical Methods in Computer Science 5*(2).

Hansson, S. O. (1990). Preference-based deontic logic (PDL). *Journal of Philosophical Logic 19*(1), 75 – 93.

Harel, D., D. Kozen, and J. Tiuryn (2000). *Dynamic Logic*. The MIT Press.

Harman, G. (1986). *Change in View: Principles of Reasoning*. The MIT Press.

References

Heifetz, A. (1996). Common belief in monotonic epistemic logic. *Mathematical Social Sciences 32*(2), 109 – 123.

Heifetz, A. (1999). Iterative and fixed point common belief. *Journal of Philosophical Logic 28*(1), 61 – 79.

Heinemann, B. (1999). Temporal aspects of the modal logic of subset spaces. *Theoretical Computer Science 224*(1-2), 135 – 155.

Heinemann, B. (2000). Extending topological nexttime logic. In S. D. Goodwin and A. Trudel (Eds.), *Temporal Representation and Reasoning*, pp. 87–94. IEEE Computer Society Press.

Heinemann, B. (2004). A hybrid logic of knowledge supporting topological reasoning. In *Algebraic Methodology and Software Technology, AMAST 2004*, Lecture Notes in Computer Science. Springer.

Holliday, W. H. (2012). *Knowing what follows: epistemic closure and epistemic logic*. Ph. D. thesis, Stanford University.

Holliday, W. H. (2014). Epistemic closure and epistemic logic I: Relevant alternatives and subjunctivism. *Journal of Philosophical Logic 44*(1), 1–62.

van der Hoek, W., B. van Linder, and J.-J. Meyer (1999). Group knowledge is not always distributed (neither is it always implicit). *Mathematical Social Sciences 38*(2), 215 – 240.

Horty, J. (2001). *Agency and Deontic Logic*. Oxford University Press.

Hughes, G. E. and M. Cresswell (1996). *A New Introduction to Modal Logic*. Routledge.

Humberstone, L. (2016). *Philosophical Applications of Modal Logic*. College Publications.

Japaridze, G. and D. de Jongh (1998). The logic of provability. In S. Buss (Ed.), *Handbook of Proof Theory*, pp. 475 – 546. Elsevier.

Jennings, R. and P. Schotch (1980). Inference and necessitation. *Journal of Philosophical Logic 9*, 327 – 340.

Jennings, R. and P. Schotch (1981). Some remarks on (weakly) weak modal logics. *Notre Dame Journal of Formal Logic 22*, 309 – 314.

de Jongh, D. and F. Liu (2009). Preference, priorities and belief. In T. Grüne-Yanoff and S. O. Hansson (Eds.), *Preference Change: Approaches from Philosophy, Economics and Psychology*, pp. 85 – 108. Springer.

Kishida, K. (2011). Neighborhood-sheaf semantics for first-order modal logic. In *Proceedings of the 7th Workshop on Methods for Modalities (M4M'2011) and the 4th Workshop on Logical Aspects of Multi-Agent Systems*, Volume 278, pp. 129 – 143.

Klein, D., N. Gratzl, and O. Roy (2015). Introspection, normality and agglomeration. In W. van der Hoek, W. H. Holliday, and W. Wang (Eds.), *Logic, Rationality, and Interaction, 5th Workshop, LORI 2015, Taipei, Taiwan*, pp. 195 – 206.

Kracht, M. (1999). *Tools and Techniques in Modal Logic*. Elsevier.

Kracht, M. and F. Wolter (1999). Normal modal logics can simulate all others. *Journal of Symbolic Logic 64*, 99 – 138.

Kratzer, A. (1977). What *must* and *can* must and can mean. *Linguistics and Philosophy 1*, 337–355.

Kremer, P. (2013). Strong completeness of S4 for any dense-in-itself metric space. *The Review of Symbolic Logic 6*, 545 – 570.

Kremer, P. (2014). Quantified modal logic on the rational line. *The Review of Symbolic Logic 7*(3), 439 – 454.

Kripke, S. (1965). Semantic analysis of modal logic II: Non-normal modal propositional calculi. In *The Theory of Models*. North Holland.

Kudinov, A. and V. Shehtman (2014). Derivational modal logics with the difference modality. In *Leo Esakia on Duality in Modal and Intuitionistic Logic*, pp. 291 – 334.

Kupke, C. and D. Pattinson (2011). Coalgebraic semantics of modal logics: An overview. *Theoretical Computer Science 412*(38), 5070 – 5094.

Lando, T. (2012). Completeness of S4 with respect to the lebesgue measure algebra based on the unit interval. *Journal of Philosophical Logic 41*(2), 287 – 316.

Lewis, D. (1969). *Convention*. Cambridge University Press.

Lewis, D. (1973). *Counterfactuals*. Oxford: Blackwell Publishers.

Lewis, D. (1974). Intensional logics without iterative axioms. *Journal of Philosophical Logic 3*, 457 – 466.

Leyton-Brown, K. and Y. Shoham (2008). *Essentials of Game Theory: A Concise, Multidisciplinary Introduction*. Morgan and Claypool Publishers.

Lismont, L. (1995). Common knowledge: Relating anti-founded situation semantics to modal logic neighbourhood semantics. *Journal of Logic, Language and Information 3*(4), 285 – 302.

Lismont, L. and P. Mongin (1994a). A non-minimal but very weak axiomatization of common belief. *Artificial Intelligence 70*(1-2), 363 – 374.

Lismont, L. and P. Mongin (1994b). On the logic of common belief and common knowledge. *Theory and Decision 37*(1), 75 – 106.

Lismont, L. and P. Mongin (2003). Strong completeness theorems for weak logics of common belief. *Journal of Philosophical Logic 32*(2), 115 – 137.

List, C. (2013). Social choice theory. In E. N. Zalta (Ed.), *The Stanford Encyclopedia of Philosophy* (Winter 2013 ed.).

List, C. (2014). Three kinds of collective attitudes. *Erkenntnis 79*(9), 1601 – 1622.

Litak, T. (2004). Modal incompleteness revisited. *Studia Logica 76*(3), 329 – 342.

Litak, T. (2005). On notions of completeness weaker than kripke completeness. In *Advances in Modal Logic*, Volume 5.

Liu, F. (2011). *Reasoning about Preference Dynamics*, Volume 354 of *Synthese Library*. Springer-Verlag.

Ma, M. and K. Sano (2015). How to update neighbourhood models. *Journal of Logic and Computation*.

Marx, M. (2007). Complexity of modal logic. In *Handbook of Modal Logic*, pp. 139 – 179. Elsevier.

McKinsey, J. and A. Tarksi (1944). The algebra of topology. *Annals of Mathematics 45*, 141 – 191.

Mints, G. and T. Zhang (2005). A proof of topological completeness for S4 in (0, 1). *Annals of Pure and Applied Logic 133*, 231 – 245.

Monderer, D. and D. Samet (1989). Approximating common knowledge with common beliefs. *Games and Economic Behavior 1*(2), 170 – 190.

Montague, R. (1970). Universal grammar. *Theoria 36*, 373 – 398.

Moss, L. and R. Parikh (1992). Topological reasoning and the logic of knowledge. In Y. Moses (Ed.), *Proceedings of TARK IV*. Morgan Kaufmann.

Moss, L., R. Parikh, and C. Steinsvold (2007). Topology and epistemic logic. In M. Aiello, I. Pratt-Hartmann, and J. van Benthem (Eds.), *Handbook of Spatial Logics*, pp. 299 – 342.

Naumov, P. (2006). On modal logic of deductive closure. *Annals of Pure and Applied Logic 141*, 218 – 224.

Negri, S. (2011). Proof theory for modal logic. *Philosophy Compass 6*, 523 – 538.

Negri, S. (2016). Proof theory for non-normal modal logics: The neighbourhood formalism and basic results. *IfCoLog Journal of Logics and their Applications*.

Pacuit, E. (2013a). Dynamic epistemic logic I: Modeling knowledge and belief. *Philosophy Compass 8*(9), 798 – 814.

Pacuit, E. (2013b). Dynamic epistemic logic II: Logics of information change. *Philosophy Compass 8*(9), 815 – 833.

Pacuit, E. and O. Roy (2015). Epistemic foundations of game theory. In E. N. Zalta (Ed.), *The Stanford Encyclopedia of Philosophy* (Spring 2015 ed.).

Pacuit, E. and S. Simon (2011). Reasoning with protocols under imperfect information. *Review of Symbolic Logic 4*(3), 412 – 444.

Palmigiano, A., S. Sourabh, and Z. Zhao (2016). Sahlqvist theory for impossible worlds. manuscript, https://arxiv.org/pdf/1603.08202.pdf.

Parikh, R. (1985). The logic of games and its applications. In M. Karpinski and J. v. Leeuwen (Eds.), *Topics in the Theory of Computation*, Annals of Discrete Mathematics 24. Elsevier.

Pattinson, D. and L. Schröder (2006). PSPACE reasoning for rank-1 modal logics. In *Proceedings of Logics in Computer Science (LICS)*, pp. 231 – 240.

Pauly, M. (1999). Bisimulation for general non-normal modal logic. Manuscript.

Pauly, M. (2001). *Logic for Social Software*. Ph. D. thesis, Institute for Logic, Language and Computation, University of Amsterdam.
Pauly, M. (2002). A modal logic for coaltional powers in games. *Journal of Logic and Computation 12*(1), 149 – 166.
Pauly, M. (2007). Axiomatizing collective judgment sets in a minimal logical language. *Synthese 158*(2), 233 – 250.
Pauly, M. and R. Parikh (2003). Game logic - an overview. *Studia Logica 75*(2), 165 – 182.
Peleg, B. (1998). Effectivity functions, game forms, games and rights. *Social Choice and Welfare 15*, 67 – 80.
Peleg, D. (1987). Concurrent dynamic logic. *Journal of the ACM 34*, 450 – 479.
Plaza, J. (1989). Logics of public communications. In M. L. Emrich, M. S. Pfeifer, M. Hadzikadic, and Z. Ras (Eds.), *Proceedings, 4th International Symposium on Methodologies for Intelligent Systems*, pp. 201–216.
Pörn, I. (1977). *Action Theory and Social Science*. Kluwer Academic Publishers.
Priest, G. (2008). *Introduction to Non-Classical Logic: From Ifs to Is*. Cambridge University Press.
Rasiowa, H. and R. Sikorski (1963). *The mathematics of metamathematics*. Państwowe Wydawnictwo Naukowe.
Roelofsen, F. (2007). Distributed knowledge. *Journal of Applied Non-Classical Logics 17*(2), 255 – 273.
Rott, H. (2001). *Change, Choice and Inference: A Study in Belief Revision and Nonmonotonic Reasoning*. Oxford University Press.
Rott, H. (2006). Shifting priorities: Simple representations for 27 iterated theory change operators. In H. Lagerlund, S. Lindström, and R. Sliwinski (Eds.), *Modality Matters: Twenty-Five Essays in Honor of Krister Segerberg*, Volume 53 of *Uppsala Philosophical Studies*, pp. 359 – 384.
Sahlqvist, H. (1975). Completeness and correspondence in the first and second order semantics for modal logic. In S. Kanger (Ed.), *Proceedings of the Third Scandinavian Logic Symposium*.
Schotch, P. and R. Jennings (1980). Modal logic and the theory of modal aggregation. *Philosophia 9*, 265 – 298.
Schröder, L. (2006). A finite model construction for coalgebraic modal logic. In *Foundations Of Software Science And Computation Structures,*, pp. 151 – 171.
Schröder, L. and D. Pattinson (2007). Rank-1 modal logics are coalgebraic. In *Proceedings of STACS*, pp. 573 – 585.
Scott, D. (1970). Advice in modal logic. In *Philosophical Problems in Logic*, pp. 143 – 173. K. Lambert.
Segerberg, K. (1971). *An Essay in Classical Modal Logic*. Filosofisska Stuier, Uppsala Universitet.
Shehtman, V. (1977). On incomplete propositional logics. *Soviet Mathematics Doklady 18*, 985 – 989.
Shehtman, V. (1980). Topological models of propositional logics. *Semiotika i informatika 15*, 74 – 98 (In Russian).
Shehtman, V. (1990). Derived sets in euclidean spaces and modal logic. Technical report, University of Amsterdam.
Shehtman, V. (2005). On neighbourhood semantics thirty years later. In *We Will Show Them! Essays in Honour of Dov Gabbay*, pp. 663 – 692. College Publications.
Spaan, E. E. (1993). *Complexity of modal logics*. Ph. D. thesis, University of Amsterdam.
Stalnaker, R. (1968). A theory of conditionals. In N. Rescher (Ed.), *Studies in Logical Theory*, pp. 98 – 112. Oxford.
Stalnaker, R. (2006). On logics of knowledge and belief. *Philosophical Studies 128*, 169 – 199.
Stalnaker, R. and R. Thomason (1970). A semantic analysis of conditional logic. *Theoria 36*, 23 – 42.
Stolpe, A. (2003). QMML: Quantified minimal modal logic and its applications. *Logic Journal of the IGPL 11*(5), 557–575.
Surendonk, T. (2001). Canonicity for intensional logics with even axioms. *The Journal of Symbolic Logic 66*(3).

Swanson, E. (2011). On the treatment of incomparability in ordering semantics and premise semantics. *Journal of Philosophical Logic 40*(6), 693 – 713.

Taylor, A. D. and W. S. Zwicker (1999). *Simple Games: Desirability Relations, Trading, Pseudoweightings*. Princeton University Press.

ten Cate, B., D. Gabelaia, and D. Sustretov (2009). Modal languages for topology: Expressivity and definability. *Annals of Pure and Applied Logic 159*(1–2), 146 – 170.

ten Cate, B. and T. Litak (2007). Topological perspective on the hybrid proof rules. *Electronic Notes in Theoretical Computer Science 174*(6), 79 – 94.

Thomason, S. K. (1972). Noncompactness in propositional modal logic. *Journal of Symbolic Logic 37*(4), 716 – 720.

Thomason, S. K. (1974). An incompleteness theorem in modal logic. *Theoria 40*, 30–34.

Vanderschraaf, P. and G. Sillari (2014). Common knowledge. In E. N. Zalta (Ed.), *The Stanford Encyclopedia of Philosophy* (Spring 2014 ed.).

Vardi, M. (1989). In *Proceedings of the 4th IEEE Symposium on Logic in Computer Science*, pp. 243 – 252.

Veltman, F. (1976). Prejudices, presuppositions and the theory of conditionals. In J. Groenendijk and M. Stokhof (Eds.), *Amsterdam Papers in Formal Grammar*, pp. 248–281. University of Amsterdam Press.

Venema, Y. (2007). Algebras and coalgebras. In J. van Benthem, P. Blackburn, and F. Wolter (Eds.), *Handbook of Modal Logic*, pp. 332 – 426.

Waagbø, G. (1992). Quantified modal logic with neighborhood semantics. *Zeitschrift für Mathematische Logik und Grundlagen der Mathematik 38*.

Wáng, Y. N. and T. Agotnes (2013). Multi-agent subset space logic. In *IJCAI 2013, Proceedings of the 23rd International Joint Conference on Artificial Intelligence*, pp. 1155 – 1161.

Wansing, H. (1998). *Displaying Modal Logic*. Kluwer Academic Publishers.

Weiss, A. and R. Parikh (2002). Completeness of certain bimodal logics of subset spaces. *Studia Logica 71*(1), 1–30.

Zvesper, J. (2010). *Playing with Information*. Ph. D. thesis, ILLC University of Amsterdam Dissertation Series DS-2010-02.

GPSR Compliance

The European Union's (EU) General Product Safety Regulation (GPSR) is a set of rules that requires consumer products to be safe and our obligations to ensure this.

If you have any concerns about our products, you can contact us on

ProductSafety@springernature.com

In case Publisher is established outside the EU, the EU authorized representative is:

Springer Nature Customer Service Center GmbH
Europaplatz 3
69115 Heidelberg, Germany

www.ingramcontent.com/pod-product-compliance
Ingram Content Group UK Ltd.
Pitfield, Milton Keynes, MK11 3LW, UK
UKHW022231230426
12048UKWH00016BA/1196